Slow Electronics with Reservoir Computing

Isao H. Inoue
Editor

Slow Electronics with Reservoir Computing

Energy-Efficient Neuromorphic Edge Computing for Low-Frequency Signals

Springer

Editor
Isao H. Inoue
National Institute of Advanced Industrial
Science and Technology (AIST)
Tsukuba, Japan

ISBN 978-981-96-8382-6 ISBN 978-981-96-8383-3 (eBook)
https://doi.org/10.1007/978-981-96-8383-3

This Springer imprint is published by the registered company Springer Nature Singapore Pte Ltd.
The registered company address is: 152 Beach Road, #21-01/04 Gateway East, Singapore 189721, Singapore

Preface

Faster is better. This has been the guiding principle of electronics for decades. From microprocessors to memory chips, from artificial intelligence to the internet, the relentless pursuit of speed has driven technological innovation. But what if, in some cases, *slower* could be the key to something even greater?

This book is an invitation to explore a radical new perspective—**Slow Electronics**. Instead of chasing ever-faster processors and higher clock speeds, Slow Electronics embraces time as a fundamental design element. Inspired by the way biological systems operate—neurons that fire over milliseconds, homeostasis that adapts over hours, ecosystems that evolve over centuries—this field seeks to build electronic devices that process information over long timescales, consuming minimal power while seamlessly integrating into our lives.

Imagine a world where devices do not constantly demand our attention but instead quietly support us in the background—learning our habits, adapting to our needs, and responding at a human pace. Wearables that monitor health not with rigid thresholds but by understanding our unique physiological rhythms. Artificial intelligence that doesn't rely on vast, power-hungry cloud networks but instead thinks locally, gently nudging us towards better choices. Machines that do not merely react, but anticipate, preserving energy and resources for a more sustainable future.

In these pages, you will find the principles, the technologies, and the possibilities of Slow Electronics. You will see how concepts from neuromorphic computing, time-domain circuits, and bio-inspired design are converging to create a new class of devices—ones that do less, but in doing so, achieve far more. This book is not just about electronics; it is about a new way of thinking about technology's role in our lives.

At its core, **Slow Electronics** is about shifting our mindset from raw speed to thoughtful efficiency, from rapid response to deep adaptation. If we embrace this paradigm, we may find ourselves living in a world where technology no longer races ahead of us, but instead walks beside us—quietly, intelligently, and with purpose.

Welcome to the future of electronics, where being slow is not a limitation, but a revolution.

Tsukuba, Japan
April 2025

Isao H. Inoue

Contents

Introduction: What Is 'Slow Electronics.'

Isao H. Inoue

Abstract This chapter introduces the concept of slow electronics, a novel paradigm in electronic device design that challenges the prevailing trend of ever-faster processing. Inspired by the ultra-low-power efficiency of biological neural networks, slow electronics aims to develop computational architectures that interact with long-time-scale environmental data in real time. The chapter explores the limitations of conventional artificial neural networks (ANNs), which struggle with high power consumption and inefficiency in processing slow-varying information. To address these challenges, the concept of neuromorphic devices–artificial neurons and synapses with extended time constants–is proposed. Additionally, the chapter discusses spiking neural networks (SNNs) and reservoir computing, which enable real-time adaptation without the computational overhead of traditional deep learning. A key feature of slow electronics is its reliance on attractor dynamics, which allow for efficient learning and inference in long-time-scale data. By integrating these three pillars–neuromorphic devices, environment-responsive architectures, and attractor-based algorithms–slow electronics paves the way for ultra-low-power, human-friendly computing. This approach has profound implications for edge computing, wearable AI, and intelligent assistive technologies, offering a sustainable alternative to cloud-based machine learning.

1 Artificial Neural Network (ANN)

The architecture known as **artificial neural network (ANN)** was developed by mimicking the structure of the biological brain [1]. The term *architecture* may sound unfamiliar, but it is a technical term that refers to the overall design philosophy and configuration of a computer system. Essentially, it describes how different components are arranged and how data is processed within the system.

I. H. Inoue (✉)
National Institute of Advanced Industrial Science and Technology (AIST), Tsukuba, Japan
e-mail: isaocaius@gmail.com

I. H. Inoue (ed.), *Slow Electronics with Reservoir Computing*,
https://doi.org/10.1007/978-981-96-8383-3_1

ANNs are often said to mimic the brain. However, they do not faithfully replicate the processes that occur within the brain. Since ANN architectures must be feasible within the constraints of modern computing, they are significantly simplified compared to biological neural networks [2, 3]. Nevertheless, ANN architectures are fundamentally different from the conventional von Neumann architecture used in most modern digital computers. Many researchers propose that ANN is the key to surpassing the limitations of von Neumann-based information processing, and consequently, ANN research has a long history as the mathematical foundation of next-generation computing [4].

Recently, ANN-based information processing has seen groundbreaking advancements, propelling ANN into an unprecedented level of prominence [5]. This is primarily due to the improvement in computing power and the availability of vast datasets, commonly referred to as big data. These factors have enabled ANN functions such as "learning" and "inference" to achieve breakthroughs far beyond what researchers had previously anticipated. Today, ANN is commonly referred to as artificial intelligence (AI).

The term "learning" in the context of ANN differs from its everyday usage. In ANN, "learning" refers to the process of adjusting numerous internal parameters to extract patterns from data. To illustrate this concept, let us consider an ANN designed to distinguish images of cats (Fig. 1). The learning process follows these steps:

1. Initially, all parameter values in the ANN are set randomly.
2. An image of a cat is converted into numerical data and input into the ANN. The ANN's output is then defined as the *teacher signal* for the cat image. (There are various methods to define teacher signals; this is just one example.)
3. Another image of a cat is input into the ANN, and the error between the output and the teacher signal is calculated.
4. The ANN adjusts its parameters to minimise this error. (This adjustment algorithm is the core of ANN learning.)
5. Another cat image is input into the ANN, and the error between the output and the teacher signal is calculated.
6. The parameters are further adjusted to minimise the error, just as in step 4.
7. By repeating steps 3 to 6 for many different cat images, the ANN gradually improves its accuracy, reducing the error between the output and the teacher signal to an extremely small value.
8. Once the error becomes consistently small for all images, the learning process is complete.

After learning, the ANN moves to the "inference" phase. However, in this context, "inference" is a technical term that refers to making predictions or judgments based on new input data. In our example, a previously unseen cat image is input into the trained ANN. If the learning was successful, the output should still yield a small error compared to the teacher signal, indicating that the system correctly identifies the image as a cat. On the other hand, if an image of a human face is input, the ANN will produce a large error, leading to the inference that the image is *not a cat*.

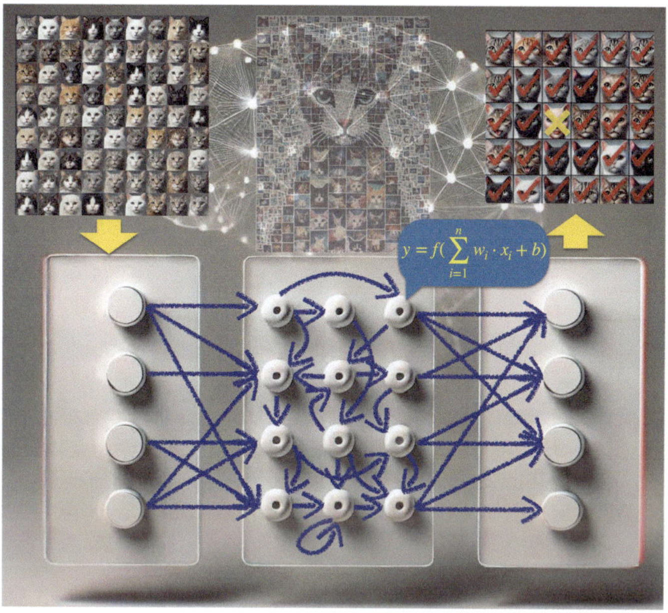

Fig. 1 Conceptual diagram of an ANN applied to image classification (cat face recognition). The input images of various cat faces (top left) are converted into pixel values and fed into the ANN's input layer (left panel). The circles in the panel represent artificial neurons. Each artificial neuron performs a computation using the activation function: $y = f\left(\sum_i w_i \cdot x_i + b\right)$, where $w_i \cdot x_i$ represents the input from the ith artificial synapse, and w_i is known as the synapse weight. Artificial synapses are depicted as blue arrows, indicating dense connectivity between artificial neurons. To enable the network to learn complex patterns, the activation function must be nonlinear. Data is propagated from the input layer to the hidden layers (middle panel). Some neural networks have structured layers, while others do not. The neurons may be systematically connected by synapses or randomly connected, as shown in this figure. The abstract representation in the upper middle of the diagram symbolises the abstraction process within the hidden layers, though it does not have a specific meaning. The output layer (right panel) compares cat and non-cat faces. Learning in ANN involves adjusting the synapse weights (w_i) so that the ANN can reliably distinguish between cat and non-cat images. If the learning process is successful, as shown in the top right, the ANN accurately classifies cat and non-cat images

Now, let us examine the information processing mechanism within ANN in more detail. ANNs consist of "artificial neurons" and "artificial synapses." Both are simply **operators that process input data**, representing a highly simplified version of the role played by neurons in the biological brain. They do not faithfully mimic the biochemical reactions occurring in biological cells.

In this section, we will focus on artificial neurons. While the neuron model introduced in Sect. 5 is quite different, artificial neurons in ANN are generally defined as operators that perform the following computation:

$$y = f\left(\sum_i w_i \cdot x_i + b\right),$$

where f is an activation function (a nonlinear transformation function), and b is a bias (adjustment value). The terms w_i and x_i are related to artificial synapses, which function as **operators that weight input signals**. Here, x_i represents data entering the artificial neuron through the ith artificial synapse, and w_i (the synapse weight) indicates the importance of that input. The *parameter values*, in the explanation of ANN above, primarily refer to these synapse weights. Thus, the learning process consists of adjusting the synapse weights in numerous artificial synapses.

ANNs are often compared to factories, where neurons act as workers who receive raw materials (inputs), process them, and send them to the next stage. Synapses serve as conveyor belts between workers, but each input signal is assigned an importance label along the belt. The overall factory layout, determining where workers are placed and how conveyor belts connect them, corresponds to the ANN architecture.

A typical ANN architecture consists of the following components (Fig. 1):

- **Input Layer**: Receives external data such as image pixel values or audio waveforms. The data is transformed into a numerical format that the ANN can process and is passed to the hidden layer.
- **Hidden Layer(s)**: Each neuron aggregates incoming data and applies a nonlinear transformation via an activation function before transmitting it to the next neuron. Synapses connecting the neurons apply weights to the data, reflecting their significance.
- **Output Layer**: Produces the final result. In the case of image recognition, the ANN outputs an error value comparing the classification result to the teacher signal.

The connectivity between neurons within and across layers is crucial. A representative example of machine learning is **deep learning (DL)**, which emerged by stacking multiple hidden layers in an ANN [6]. DL employs sophisticated connection methods such as convolution [7]. In contrast, **reservoir computing**, a different type of ANN, does not use multiple layers in its hidden structure [8, 9]. In reservoir computing, synapse weights are not adjusted in the hidden layers; instead, only the output layer is modified. Chapter 2 will provide further details on reservoir computing.

2 The Nobel Prize in Physics for ANN

The 2024 Nobel Prize in Physics was awarded for research on artificial neural networks (ANN) [10]. This news broke while I was writing this book, and it created a huge stir in the media. On social media, the announcement sparked a heated debate, with both praise and criticism being voiced. As a researcher specialising in condensed matter physics, I had predicted that the 2024 Nobel Prize in Physics might be awarded for neural networks. I even shared my prediction on social media in advance. However, when it turned out to be true, I was so surprised that I could in fact hardly believe it. Nevertheless, given the significance of this event, I feel compelled to address it in this chapter.

The decision to award the Nobel Prize to ANN research is deeply rooted in the *close relationship between physics and artificial neural networks*. In the previous section, I mentioned that ANN can be used to recognise images of cats. But if we think more deeply about why this is possible, we realise that physics plays a fundamental role.

First, the learning process in ANN is based on principles from physics, such as energy minimisation and statistical mechanics [11]. In the section on ANN "learning," I explained that parameters are adjusted to minimise errors. This minimisation process is often carried out using algorithms like stochastic gradient descent, which is based on statistical-mechanics techniques used to analyse system behaviour [12]. It closely resembles the principle that natural systems tend to evolve towards states of lower energy [13].

In essence, the way neurons are interconnected and how synaptic weights are adjusted can be seen as *modelling complex interactions*–something that lies at the very heart of physics research [14]. The multi-layered structure and non-linearity of ANN are also crucial for their functionality. These aspects are closely linked to complex-systems physics and network science [6]. From this perspective, ANN research and physics share a common goal: understanding the intricate interactions in nature.

ANN is no longer just a tool for information processing; it is increasingly regarded as a subject of study that offers insights through a physics-based approach. In fact, ANN has recently been applied in various areas of physics. In quantum physics, it is used for optimising quantum states and simulations [15]. In astrophysics, it helps analyse astronomical images [16]. In fluid dynamics, it is being utilised for simulations [17]. ANN is driving innovation in both theoretical and experimental physics.

Given these remarkable contributions, it is fair to say that ANN research was truly deserving of the Nobel Prize.

3 The Rise of Deep Learning

Many applications of artificial neural networks (ANNs) in physics are based on deep learning (DL). DL has demonstrated the ability to efficiently solve complex information processing problems, such as image recognition, speech recognition, and natural language processing, which were previously considered highly challenging. This has led to the establishment of a new paradigm [6]. Beyond physics research, DL is widely used across many aspects of modern society, and the impact of its technological foundation is profound.

In recent years, a new DL architecture known as the Transformer has gained significant attention [18]. Transformers are based on a mechanism called *self-attention*, which allows the model to consider the entire input data at once and efficiently learn which parts are interrelated [19]. Unlike conventional convolutional neural networks (CNNs) and long short-term memory networks (LSTMs), which process data sequentially through intermediate layers, Transformers eliminate this need. This enables much faster and more accurate learning, paving the way for the development of so-called *generative AI* [20].

Traditional DL models are often referred to as *discriminative AI* because they excel at classifying data and assigning labels. In contrast, generative AI can create new content from data. This capability has led to remarkable advancements in areas such as text generation and translation in natural language processing, as well as the generation of images and illustrations in image processing [21]. It is no exaggeration to say that modern societal progress is now inseparable from DL.

The rapid advancement of DL has been driven by the widespread adoption of the internet, which has enabled easy access to vast amounts of data, commonly referred to as *big data*. In modern society, a variety of information is transmitted via the internet to data centres, often referred to as the *cloud*. Data centres are large-scale facilities that house extensive server clusters, providing computational resources for learning and inference tasks. As mentioned earlier, learning in DL involves operations such as weighting, summation, and the application of activation functions at the level of synapses and neurons. Since all these calculations occur in parallel across synapses and neurons, the computational load is enormous [22]. However, with the rapid improvement in computing power, performing *massive computations* on *massive datasets* is no longer an impossibility, which has driven the remarkable progress of DL in recent years.

During training, DL models process large amounts of data simultaneously through a method known as *batch processing*. However, since the number of ANN parameters (weights) ranges from millions to billions, these must be stored in memory, read, updated, and saved repeatedly–an unimaginably large number of times [19]. This process consumes significant energy, particularly in data centres, where dynamic random-access memory (DRAM) operates continuously to handle vast amounts of data at high speed. In fact, DRAM usage is one of the biggest contributors to power consumption in these facilities.

For example, Google's Bidirectional Encoder Representations from Transformers (BERT), which achieved groundbreaking results in natural language processing (NLP), was trained using four GPUs on the Wiki+Books dataset with one million steps of masked language modelling. This process took approximately 726, 553 seconds and consumed a staggering 146.15 kWh of energy [23]. OpenAI's GPT-3, with its 175 billion parameters, requires an estimated 355 years and 4.6 million USD if trained using Tesla V100 GPUs on a cloud platform at a cost of 1.50 USD per hour per unit, with eight units running simultaneously [24].

Today, data centres are installed worldwide, and their total energy consumption accounts for approximately $1 \sim 2\,\%$ of global electricity usage [25]. This proportion is increasing annually, with the growing adoption of generative AI driving a sharp rise in DL-related power demands [26]. In some regions where large-scale data centres are located, the increasing electricity demand has placed stress on other infrastructure. It may even affect regional power supply and energy pricing, necessitating urgent countermeasures. As a response, the development of *DL-specific chips*, such as Google's TPU, is being actively pursued [27]. These chips are designed specifically for DL training and offer higher energy efficiency than conventional GPUs. However, it remains unclear how much the introduction of DL-specific chips can mitigate the growing energy demand, making the long-term outlook uncertain [28].

The human brain can be regarded as an "ultra-low-power computer" that operates on just 20 W of energy [29, 30]. Ironically, however, the biggest challenge facing ANN, which was originally designed to mimic the brain, is its immense power consumption. This is because current ANNs only superficially replicate the functions of biological neurons and synapses, rather than truly mimicking the brain's overall information processing capabilities. As a result, research is now emerging to develop ANNs that more closely resemble the brain [31, 32]. The **slow electronics** approach introduced in this book is one such effort.

4 Edge Devices That Support People

Deep learning (DL) has advanced rapidly, and we are now at a point where it can assist us in making complex decisions in daily life. This could significantly improve our quality of life. Consider the devices we use every day–smartphones, smartwatches, spectacles, and accessories. Now, imagine more advanced wearable and implantable technologies, such as contact lenses that monitor eye health, implantable cardioverter-defibrillators (ICDs) that regulate heartbeats, continuous glucose monitors (CGMs) for diabetics, and bone-anchored hearing aids (BAHAs) that improve hearing. These devices are becoming increasingly capable of observing human behaviour and physiological signals. Through machine learning, they can identify patterns, detect potential health risks, and even take action when something seems wrong [33, 34]. Most bodily functions are regulated by the autonomic nervous system, meaning they follow *natural rhythms and patterns*. Machine learning excels at recognising such regularities, making it an ideal tool for this purpose.

However, a truly supportive edge device should do more than detect standard patterns. It must learn and adapt to an individual's unique traits and the subtle changes that occur over time. These fluctuations–often difficult to define–make up a person's habits, instincts, and even artistic sensibilities. Imagine a device that could "visualise" an artist's creativity or a craftsman's technique, which are difficult to quantify mathematically. Such technology could help preserve and pass down traditional arts and skills while also fostering innovation. It would not only document techniques but also reveal the deeper essence of craftsmanship.

A major drawback of current DL-based machine learning is its enormous power consumption. Learning individual traits such as artistic habits or personal preferences requires intimate, continuous adaptation. However, sending all data to cloud-based supercomputers–where processing power is virtually unlimited–comes with significant privacy and security risks [35]. One alternative is to use pre-trained models that only perform inference at the edge [36]. While this reduces power consumption, it does not allow *real-time learning of an individual's unique characteristics*, which is crucial for **devices meant to support people**.

The concept of connecting everything to the internet and leveraging cloud-based DL is at the core of the Internet of Things (IoT) [37]. However, as shown in Fig. 2, IoT poses significant challenges for learning and inference based on private data–an essential feature of truly supportive devices. For such devices to work effectively, they must be able to *learn and infer in real time without relying on cloud systems*. Edge devices that process data locally can maintain privacy while still learning and adapting.

One particularly exciting innovation is the ability to *visualise personal traits*–such as creativity, instincts, and preferences–without compromising privacy. This technology could fundamentally change our lives. By making success and failure patterns visible only to the individual, it could support self-reliance among the elderly, assist people with disabilities, and enhance skills in a way that is both personalised

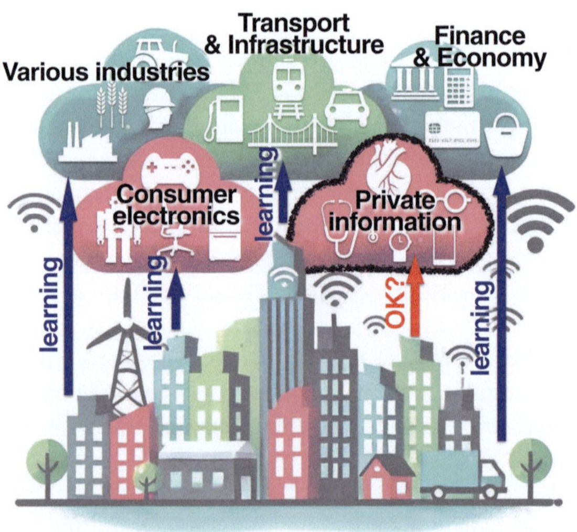

Fig. 2 Modern devices and sensors continuously send data to cloud-based supercomputers, as illustrated by the upward arrow labeled "learning" in this figure. The cloud processes this data using machine learning and provides feedback to users. This framework, known as IoT, underpins many sectors, including: agriculture, manufacturing, and energy; transportation and infrastructure (e.g., vehicles, bridges, fuel stations); finance and economics (e.g., banking, digital payments, economic monitoring); consumer electronics (e.g., smart appliances, personal devices, office equipment); and even personal information (e.g., health data, daily habits). While IoT enables powerful data-driven learning, it also raises concerns. How is personal data handled? Are privacy protections sufficient? These concerns are especially critical in fields such as healthcare and behavioural monitoring

and non-intrusive. Additionally, real-time learning would allow a device to *predict the next step* for its user. This kind of self-awareness could significantly improve one's quality of life.

Behavioural economics suggests that frequent feedback and small, immediate rewards are highly effective in encouraging behavioural change [38]. Devices that provide real-time feedback could leverage these insights to help users develop healthier habits, improve daily routines, and even break harmful cycles such as smoking or excessive drinking. Ultimately, these devices will do more than process information. They will support the human mind, helping individuals stay mindful, ethical, and connected in an increasingly digital world.

5 Spiking Neural Network (SNN)

The human brain is remarkably efficient in processing information in its environment, consuming only 20 W of energy [29, 30]. It goes without saying that the brain represents the ultimate form of a "human-friendly device." This information-processing system known as the brain is composed of two fundamental components: neurons and synapses. Both of these components operate at a slow pace and exhibit significant variability. Moreover, synapses exhibit a hysteresis effect, a phenomenon that would typically be considered a defect in conventional semiconductor devices. However, life forms have been adapting and optimising these "slow and unreliable components" for over 500 million years [39], ultimately developing the brain–a hardware system capable of flexible learning and inference with extremely low power consumption. The brain is not only a human-friendly device but is also regarded as an ideal model for electronics [40].

One of the core functions of biological neurons is their ability to process input signals dynamically. Neurons possess an internal parameter known as the **membrane potential**, which varies in response to external input signals (Fig. 3). These input signals take the form of electrical pulses known as **spikes**, which serve as the means of communication between neurons in the brain. If incoming spikes arrive at a frequency higher than the time constant of membrane potential changes, the membrane potential accumulates. Once it reaches a certain threshold, the neuron emits a spike–a phenomenon known as neuronal **firing**. Conversely, if the frequency of incoming signals is too low, the membrane potential does not accumulate sufficiently, and firing does not occur. This property is captured by a simple model known as the **leaky integrate-and-fire (LIF)** model [41]. An important feature of biological neurons is the presence of **leak**, which causes the membrane potential to decrease gradually over time if no further spikes arrive. This decay ensures that neurons do not indefinitely retain past inputs but instead respond dynamically to new stimuli. The combination of spikes and leak mechanisms contributes significantly to the brain's efficiency as a low-power information-processing system [42].

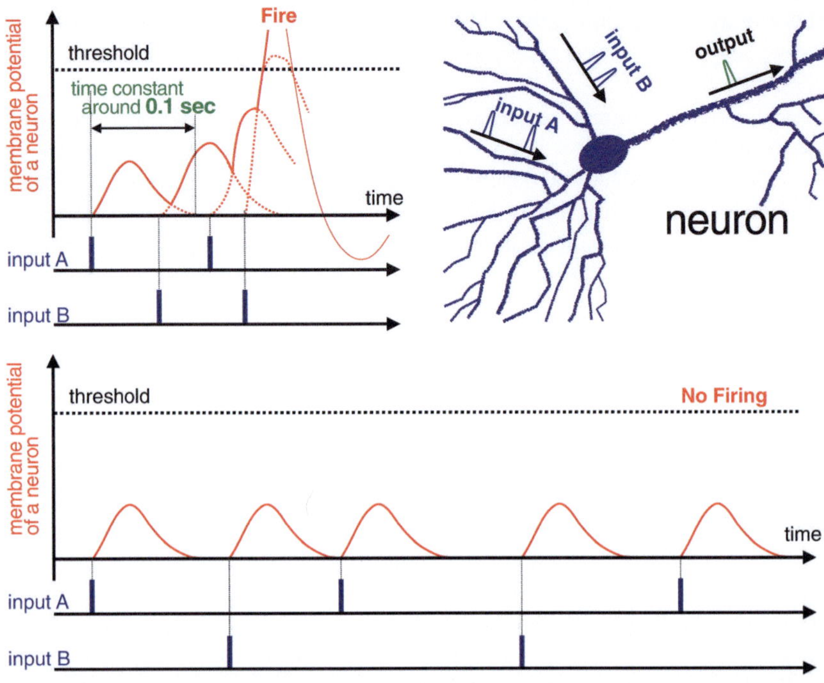

Fig. 3 The relationship between the timescale of external inputs and neuronal firing in the LIF model of biological neurons (top right). The top left panel illustrates a scenario where the timescale of external inputs (Input A and Input B) is shorter than the neuron's time constant (approximately 0.1 s). When an external input is present, the membrane potential inside the neuron increases and then decreases with a time constant of 0.1 s. If another input arrives before the membrane potential returns to zero, the potential accumulates. In this case, if the membrane potential exceeds the firing threshold, a spike is generated. The bottom panel shows a scenario where the timescale of external inputs is longer than the neuron's time constant. Since the next input arrives only after the membrane potential has returned to zero, the potential never reaches the firing threshold, and no spike occurs. The LIF model embodies a key principle of neuronal dynamics: neurons with short time constants are necessary for responding to rapid and transient stimuli, whereas neurons with long time constants are required for processing slow and sustained inputs effectively enough to reach the firing threshold

Traditional artificial neural networks (ANNs), as described in Fig. 1, use mathematical models such as $y = f\left(\sum w_i \cdot x_i + b\right)$ to process information. Unlike biological neurons, artificial neurons in ANNs do not generate spikes and lack the leak mechanism, making them fundamentally different from how the brain operates. To bridge this gap, **spiking neural networks (SNNs)** have been developed [43]. SNNs incorporate the principles of biological neurons by using spikes as their fundamental mode of communication. By mimicking the LIF model, SNNs process information more efficiently, particularly in scenarios where low-power computation and event-driven processing are crucial. This book focuses on machine learning based on these SNNs.

6 The Necessity of Slow Electronics

Utilising SNNs to achieve ultra-low-power, brain-inspired information processing is considered a conventional approach in the development of **human-friendly devices**. However, this method faces a critical challenge: in time-series data derived from living environments, *many key features emerge on long time scales*, typically ranging from $\sim 10\,\text{ms}$ to $\sim 10\,\text{s}$ [44] (Fig. 4). One might wonder why this is such a critical issue. In fact, in order to utilise time-series data for real-time inference through machine learning, it is essential to match the time scale of the data with that of the neural network. As discussed in this paper [45] and detailed in Chapter 3, when employing reservoir computing using SNNs for handwriting-based personal authentication, successful authentication occurs when the neuron time constant is 1.07 s. However, authentication fails when the time constant is as short as $10\,\mu\text{s}$. This suggests that if the neuron time constant is sufficiently long compared to the time scale of the characteristic features within the handwriting data, machine learning can effectively capture those features. On the other hand, if the neuron time constant is too short, as

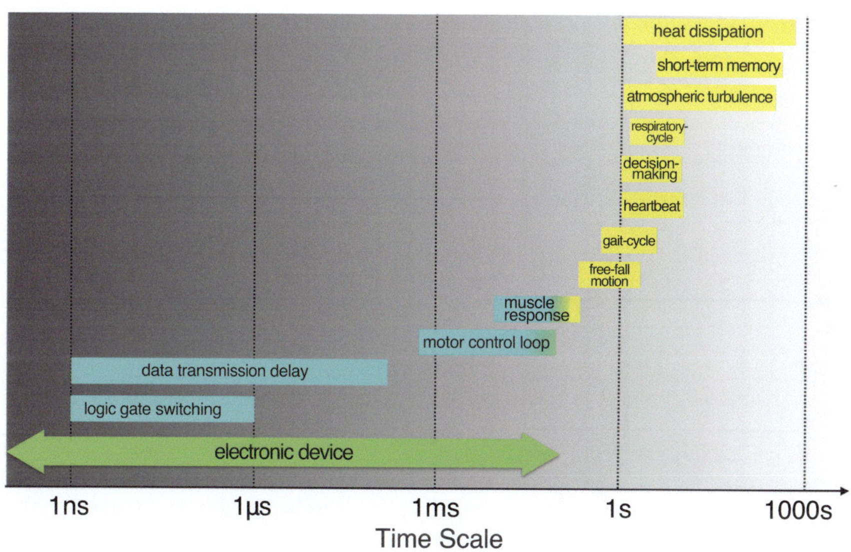

Fig. 4 A visual representation of characteristic time scales associated with various phenomena observed in human living environments. The temporal dynamics of electronic devices, physiological processes, cognition, and physical phenomena differ significantly. Examples where the time scale aligns with that of electronic components (green arrows) are shown in blue bars, making them suitable for machine learning using ANNs on conventional digital computers. In contrast, phenomena related to physiological or natural processes exhibit much longer characteristic time scales, as indicated by the yellow bars. Achieving energy-efficient real-time machine learning at the edge requires matching the time scale of the phenomenon with that of the neural network. However, this alignment is extremely difficult with existing digital ANN implementations

shown in the lower panel of Fig. 3, learning becomes impossible. While 10 μs may seem like a relatively long time constant in comparison to modern semiconductor devices, it remains far too short for real-time learning of everyday phenomena such as handwriting.

Modern high-speed digital computers are not incapable of handling time-series data with long time scales. By appropriately storing input data, compressing it in the time axis, and processing it in batches, such data can be managed (Fig. 5). However, the focus of this book is real-time machine learning at the edge. Batch processing compromises real-time capabilities. Furthermore, machine learning relies on neural networks with an enormous number of parameters, which exacerbates the problem. Each batch process requires loading and unloading vast internal states from non-volatile memory such as DRAM, which results in excessive power consumption [46]. Consequently, the reliance on digital computing for processing long time-scale time-series data leads to significant power usage due to memory access, making real-time machine learning at the edge highly impractical. It is no exaggeration to state that the fundamental challenge in developing **human-friendly devices** is addressing the **issue of slow data**. While learning from long time-scale phenomena may seem straightforward, it is an unexpectedly difficult problem, and one that is not well understood. In reality, this presents an essential and profound challenge.

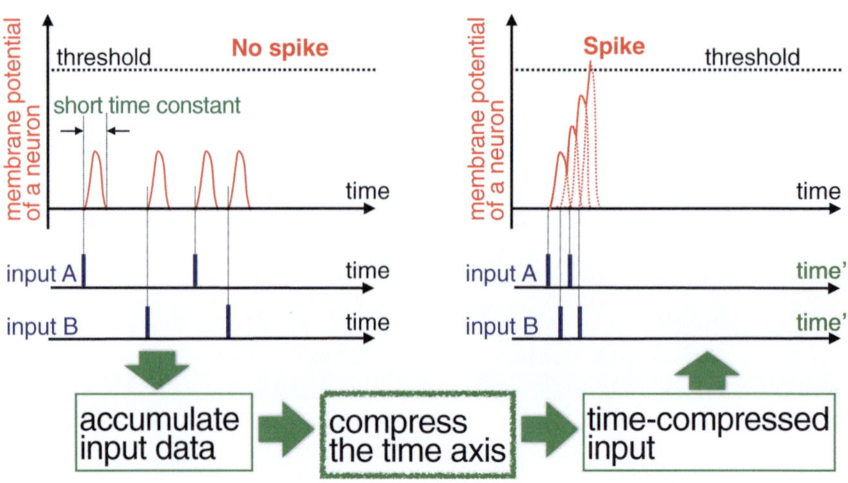

Fig. 5 When the time scale of external input is longer than the neuron time constant, the membrane potential decays before accumulating sufficiently to reach the firing threshold, preventing neuron spikes from occurring. To address slow external inputs using fast neurons, it is necessary to accumulate external input data, compress it in the time axis, and then input it into the neurons–this is known as batch processing. While batch processing allows the membrane potential to exceed the firing threshold, generating spikes, it prevents real-time processing of external inputs. Additionally, batch processing necessitates the constant retrieval and storage of large amounts of data and parameters in memory, leading to significant power consumption

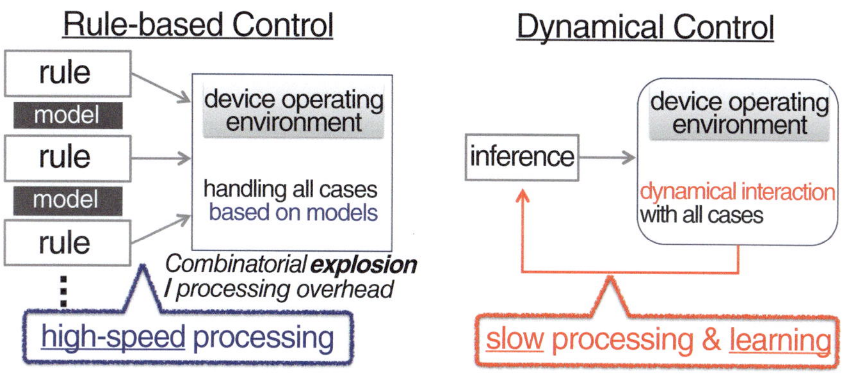

Fig. 6 A comparison between conventional rule-based control and brain-inspired dynamic control in electronic systems. Rule-based control requires predefined rules and models for all possible situations, leading to an intractable explosion of rule combinations and overwhelming processing loads. In contrast, biological brains dynamically interact with their environment, continuously learning and inferring without relying on fixed rules or models. While this process necessitates slow information processing, it eliminates the combinatorial explosion problem inherent in rule-based control, resulting in a simpler and more efficient control system

Modern electronic control systems predominantly rely on rule-based frameworks [47]. However, attempting to predefine rules for all possible events inevitably encounters the so-called combinatorial explosion–a problem classified as *NP-complete* in computational theory and referred to as the *frame problem* in artificial intelligence research (Fig. 6, left). As the scale and complexity of a system increase, the required computational time grows exponentially, resulting in an immense processing overhead [48]. A famous analogy illustrating this problem is John Searle's "Chinese Room" thought experiment [49]. Imagine a person who only understands English but is placed inside a room with an extensive manual containing responses to all possible questions in both English and Chinese. When presented with a question written in Chinese, the person follows the manual to respond in Chinese. Over time, an external observer may be misled into believing that the person inside the room understands Chinese. However, if such a perfect manual were to exist, it would be astronomically large. Moreover, the concept of anticipating "all possible questions" contradicts itself–if an unforeseen question is expected, it is no longer unforeseen. This highlights the fundamental impossibility of pre-programming all scenarios into an electronic control system, as such an approach would lead to unmanageable computational overhead. Interestingly, the human brain does not suffer from the combinatorial explosion problem. We use language, yet our brains do not store an astronomical amount of predefined dictionary data. Instead of depending on predefined rules, the brain continuously learns from interactions with the external environment, allowing it to infer dynamically and adapt to new situations (Fig. 6, right). This **ability to dynamically interact with the external environment** stands in stark contrast to the fixed, pre-programmed nature of conventional von Neumann-type electronic control systems. The adaptability of neural networks provides a foundation for learning and inference-based information processing.

It is crucial to highlight once again that biological neurons and synapses operate much more slowly than electronic circuits. This slowness is precisely what enables the brain, as a network of such slow components, to interact appropriately with long time-scale external environments. This property has given rise to an entirely different, "brain-like" information processing system, distinct from the von Neumann architecture. Although numerous studies have attempted to unravel the mysteries of brain information processing, very few have focused on the fact that neurons and synapses are slow components. Since the brain has evolved in response to its environment, it is natural that neurons and synapses have become slow. However, the prevailing trend in modern semiconductor electronics is to pursue ever-faster operation, making it difficult to conceive an approach that intentionally embraces slowness. From this perspective, this book proposes **Slow Electronics** as a novel paradigm in electronics. The goal of Slow Electronics research is to pave the way for the development of brain-inspired computing. This represents a completely unique approach that runs *counter to the prevailing trend of developing faster devices and circuits* for machine learning. By embracing slow processing, we believe this research has the potential to **drive a paradigm shift** in information processing.

7 Three Key Pillars of Slow Electronics

This book introduces the concept of **devices that stay close to people**. These devices are designed to observe time-series data that exhibit stationarity and periodicity, similar to physiological phenomena, in real time at the edge, where power supply is limited, and enable machine learning. Ultimately, such devices could help visualise characteristics that are difficult to express mathematically, such as sensibility and habits. Since these devices handle privacy-sensitive data, they cannot rely on internet connectivity for cloud-based machine learning. Moreover, most of the relevant data have distinct patterns that emerge over long time scales. Therefore, a device capable of performing *machine learning on long-time-scale data while consuming ultra-low power* is necessary–this is what we call a device that stays close to people.

The technology required to build such a device is **slow electronics**. To realise slow electronics, we propose the development of three key pillars (Fig. 7):

1. **Neuromorphic Devices** Artificial neurons and artificial synapses with long time constants, allowing them to process long-time-scale time-series data.
2. **Architectures that Interact with the Environment** Spiking neural networks (SNNs) that can dynamically interact with their environment in real time through learning and inference. This book focuses on a type of SNN called reservoirs.
3. **Utilisation of Attractors** Algorithms that can identify periodicity and deterministic structures in real-time time-series data, enabling effective learning and inference.

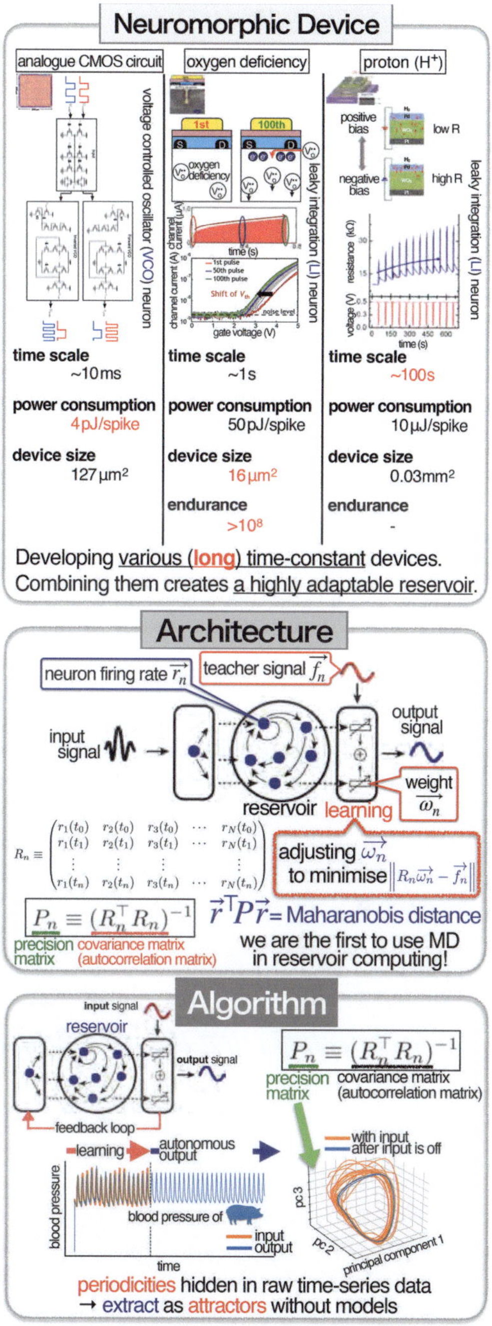

Fig. 7 The three key pillars of slow electronics. Selected examples introduced in this book are shown. Top: Examples of neuromorphic devices with long time constants, including artificial neurons utilising analogue CMOS circuits [51], oxygen vacancies [45], and protons [44, 52], each exhibiting time constants differing by a factor of approximately 100. Detailed explanations are provided in Chapter 3 and Chapter 4. Middle: Architecture for reservoir computing that interacts with the environment [9], covered in Chapter 2. Bottom: Algorithms using attractors for time-series data analysis, incorporating reservoirs with feedback loops [53]. By combining these three key pillars of slow electronics, it is possible to create ultra-low-power **devices that stay close to people**

7.1 Neuromorphic Devices

Current artificial neurons operate at relatively fast speeds, making them unsuitable for handling slow data from human environments. As illustrated in Fig. 4, very few artificial neurons with time constants exceeding 1s have been successfully developed [45]. Handling slow time-series data with conventional high-speed digital electronics is inefficient; it results in excessive power consumption and loss of real-time responsiveness (Fig. 5). Thus, the development of neuromorphic devices that can naturally process slow signals is crucial for slow electronics. Traditional analogue circuits rely on capacitors storing charge that leaks through resistors, resulting in a time constant of $\tau = RC$. However, on a silicon chip, increasing R by lengthening the wiring is limited, reaching at most a few tens of kΩ. (Interestingly, we have devised a method to utilise a pseudo-resistor, which enables the formation of a virtually large electrical resistance. By leveraging this approach, we have successfully implemented long time constants in analogue CMOS circuits [50]. However, this did not dramatically improve energy consumption.) Similarly, increasing C to extend the time constant beyond 1 s would require capacitors of impractical sizes (over 1 cm^2), making integration unfeasible.

To overcome these limitations, this book introduces three novel artificial neuron designs (Fig. 7, top):

- Artificial neurons using **voltage-controlled oscillators** in analogue CMOS circuits [51] (Chap. 4).
- Artificial neurons utilising the **electric drift** of positively charged **oxygen vacancies** and their **thermal diffusion** to achieve slow leaky integration [45] (Chap. 3).
- Artificial neurons using positively charged **protons** in oxides for even longer-time-scale leaky integration, leveraging **electric drift** and **thermal diffusion** [44, 52] (Chap. 3).

7.2 Architectures that Interact with the Environment

Modern DL models excel at processing vast datasets in large computers through pre-training before inference. However, such models are not well-suited for responding to long-time-scale time-series data in real-world human environments–an essential function for devices that stay close to people. This book proposes the use of **reservoir computing** [8, 9], which enables spiking neural networks (SNNs) to dynamically interact with their surroundings. Reservoir computing reduces computational complexity during learning and significantly improves energy efficiency. Replacing conventional fast neurons in reservoirs with slow neuromorphic devices allows real-time processing of long-time-scale data. More details on reservoir computing are discussed in Chap. 2.

7.3 *Utilisation of Attractors*

The human brain efficiently processes information using nonlinear dynamics, particularly in memory formation and decision-making [54]. A fundamental concept in this process is the **attractor**–mathematically defined as a state or trajectory that a system naturally converges to over time [11, 55, 56]. Applying attractor dynamics to slow electronics enables efficient learning and inference while minimising energy consumption. Short-term memory in the brain temporarily holds information, while long-term memory stabilises and retains it over extended periods. Different learning rules govern these processes, with the hippocampus handling short-term retention and the cortex storing long-term knowledge [57–59]. The brain retrieves memories by searching through its attractor structures based on contextual cues. Inspired by this mechanism, we are exploring new learning algorithms that *integrate reservoir computing with attractor theory* [53].

Since real-time time-series data often contain periodic or deterministic structures, incorporating attractor dynamics into reservoir computing enables efficient **feature extraction** with minimal power consumption. By applying this approach to memory and inference, we can develop highly energy-efficient intelligent devices. Chapter 7 provides a detailed discussion on these theoretical frameworks and their practical applications in slow electronics.

8 Conclusion

The Chap. 1 introduced the concept of slow electronics, explained why it is necessary, and outlined the three key pillars required for its realisation. The following chapters will explore methodologies, fabrication techniques, and real-world applications in greater detail, establishing the foundation for this emerging field.

Competing Interests

This study was funded by Japan Science and Technology Agency (JST) CREST Grant No. JPMJCR19K.

The author has no conflicts of interest to declare that are relevant to the content of this chapter.

References

1. S. Schmidgall, R. Ziaei, J. Achterberg, L. Kirsch, S.P. Hajiseyedrazi, J. Eshraghian, APL Mach. Learn. **2**, 021501 (2024). https://doi.org/10.1063/5.0186054
2. F. Shao, Z. Shen, Front. Psychol. **13**, 970214 (2022). https://doi.org/10.3389/fpsyg.2022.970214
3. S. Zheng, L. Qian, P. Li, C. He, X. Qin, X. Li, arXiv [cs.NE] (2022). https://doi.org/10.48550/arXiv.2204.07519

4. H.T. Ünal, F. Başçiftçi, Artif. Intell. Rev. **55**, 1723 (2022). https://doi.org/10.1007/s10462-021-10049-5
5. I.H. Sarker, S.N. Comput, Sci. **2**, 420 (2021). https://doi.org/10.1007/s42979-021-00815-1
6. Y. LeCun, Y. Bengio, G. Hinton, Nature **521**, 436 (2015). https://doi.org/10.1038/nature14539
7. Z. Li, F. Liu, W. Yang, S. Peng, J. Zhou, IEEE Trans. Neural Netw. Learn. Syst. **33**, 6999 (2022). https://doi.org/10.1109/TNNLS.2021.3084827
8. H. Jaeger, H. Haas, Science **304**, 78 (2004). https://doi.org/10.1126/science.1091277
9. G. Tanaka, T. Yamane, J.B. Héroux, R. Nakane, N. Kanazawa, S. Takeda, H. Numata, D. Nakano, A. Hirose, Neural Netw. **115**, 100 (2019). https://doi.org/10.1016/j.neunet.2019.03.005
10. The Nobel Prize in Physics 2024. NobelPrize.org. Nobel Prize Outreach AB 2024. https://www.nobelprize.org/prizes/physics/2024/summary/
11. J.J. Hopfield, Proc. Natl. Acad. Sci. U. S. A. **79**, 2554 (1982). https://doi.org/10.1073/pnas.79.8.2554
12. H. Robbins, S. Monro, Ann. Math. Stat. **22**, 400 (1951). https://doi.org/10.1214/aoms/1177729586
13. D.E. Rumelhart, G.E. Hinton, R.J. Williams, Nature **323**, 533 (1986). https://doi.org/10.1038/323533a0
14. P. Battaglia, R. Pascanu, M. Lai, D.J. Rezende, K. Kavukcuoglu, Neural Inf. Process Syst. **29**, 4502 (2016)
15. G. Carleo, M. Troyer, Science **355**, 602 (2017). https://doi.org/10.1126/science.aag2302
16. K. Sharma, A. Kembhavi, A. Kembhavi, T. Sivarani, S. Abraham, K. Vaghmare, Mon. Not. R. Astron. Soc. **491**, 2280 (2020). https://doi.org/10.1093/mnras/stz3100
17. M. Raissi, P. Perdikaris, G.E. Karniadakis, J. Comput. Phys. **378**, 686 (2019). https://doi.org/10.1016/j.jcp.2018.10.045
18. A. Vaswani, N. Shazeer, N. Parmar, J. Uszkoreit, L. Jones, A.N. Gomez, L. Kaiser, I. Polosukhin, arXiv [cs.CL] (2017). https://doi.org/10.48550/arXiv.1706.03762
19. A. Khan, A. Sohail, U. Zahoora, A.S. Qureshi, Artif. Intell. Rev. **53**, 5455 (2020). https://doi.org/10.1007/s10462-020-09825-6
20. T.B. Brown, B. Mann, N. Ryder, M. Subbiah, J. Kaplan, P. Dhariwal, A. Neelakantan, P. Shyam, G. Sastry, A. Askell, S. Agarwal, A. Herbert-Voss, G. Krueger, T. Henighan, R. Child, A. Ramesh, D.M. Ziegler, J. Wu, C. Winter, C. Hesse, M. Chen, E. Sigler, M.T. Litwin, S. Gray, B. Chess, J. Clark, C. Berner, S. McCandlish, A. Radford, I. Sutskever, D. Amodei, in *Advances in Neural Information Processing Systems*, vol. 33, ed. by H. Larochelle, M. Ranzato, R. Hadsell, M.F. Balcan, H. Lin (Curran Associates, Inc., 2020), vol. 33, pp. 1877–1901
21. S.S. Sengar, A.B. Hasan, S. Kumar, F. Carroll, Multimedia Tools Appl. (2024). https://doi.org/10.1007/s11042-024-20016-1
22. X. Hu, L. Chu, J. Pei, W. Liu, J. Bian, Knowl. Inf. Syst. **63**, 2585 (2021). https://doi.org/10.1007/s10115-021-01605-0
23. X. Wang, C. Na, E. Strubell, S. Friedler, S. Luccioni, in *The 2023 Conference on Empirical Methods in Natural Language Processing* (2023)
24. C. Li. OpenAI's GPT-3 language model: A technical overview (2020). https://lambdalabs.com/blog/demystifying-gpt-3
25. E. Masanet, A. Shehabi, N. Lei, S. Smith, J. Koomey, Science **367**, 984 (2020). https://doi.org/10.1126/science.aba3758
26. D.N. Forbes. Can Innovation Curb AI's Hunger for Power?, The Wall Street Journal. (2024). https://www.wsj.com/articles/can-innovation-curb-ais-hunger-for-power-e9c3d8bc
27. N.P. Jouppi, C. Young, N. Patil, D. Patterson, G. Agrawal, R. Bajwa, S. Bates, S. Bhatia, N. Boden, A. Borchers, R. Boyle, P.L. Cantin, C. Chao, C. Clark, J. Coriell, M. Daley, M. Dau, J. Dean, B. Gelb, T.V. Ghaemmaghami, R. Gottipati, W. Gulland, R. Hagmann, C.R. Ho, D. Hogberg, J. Hu, R. Hundt, D. Hurt, J. Ibarz, A. Jaffey, A. Jaworski, A. Kaplan, H. Khaitan, D. Killebrew, A. Koch, N. Kumar, S. Lacy, J. Laudon, J. Law, D. Le, C. Leary, Z. Liu, K. Lucke, A. Lundin, G. MacKean, A. Maggiore, M. Mahony, K. Miller, R. Nagarajan, R. Narayanaswami, R. Ni, K. Nix, T. Norrie, M. Omernick, N. Penukonda, A. Phelps, J. Ross, M. Ross, A. Salek,

E. Samadiani, C. Severn, G. Sizikov, M. Snelham, J. Souter, D. Steinberg, A. Swing, M. Tan, G. Thorson, B. Tian, H. Toma, E. Tuttle, V. Vasudevan, R. Walter, W. Wang, E. Wilcox, D.H. Yoon, in *Proceedings of the 44th Annual International Symposium on Computer Architecture* (ACM, New York, NY, USA, 2017). https://doi.org/10.1145/3079856.3080246

28. J.L. Hennessy, D.A. Patterson, Commun. ACM **62**, 48 (2019). https://doi.org/10.1145/3282307
29. D. Attwell, S.B. Laughlin, J. Cereb. Blood Flow Metab. **21**, 1133 (2001). https://doi.org/10.1097/00004647-200110000-00001
30. B. Sengupta, M.B. Stemmler, Proc. IEEE Inst. Electr. Electron. Eng. **102**, 738 (2014). https://doi.org/10.1109/jproc.2014.2307755
31. K. Katyal, J. Parent, B. Alicea, arXiv [cs.NE] (2021). https://doi.org/10.48550/arXiv.2103.15553
32. Y. Cohen, T.A. Engel, C. Langdon, G.W. Lindsay, T. Ott, M.A.K. Peters, J.M. Shine, V. Breton-Provencher, S. Ramaswamy, J. Neurosci. **42**, 8514 (2022). https://doi.org/10.1523/JNEUROSCI.1503-22.2022
33. B. Rush, L.A. Celi, D.J. Stone, J. Clin. Monit. Comput. **33**, 887 (2019). https://doi.org/10.1007/s10877-018-0219-z
34. Y. Qiao, J. Luo, T. Cui, H. Liu, H. Tang, Y. Zeng, C. Liu, Y. Li, J. Jian, J. Wu, H. Tian, Y. Yang, T.L. Ren, J. Zhou, Nanomicro Lett. **15**, 66 (2023). https://doi.org/10.1007/s40820-023-01029-1
35. M. Mukherjee, R. Matam, C.X. Mavromoustakis, H. Jiang, G. Mastorakis, M. Guo, I.E.E.E. Commun, Mag. **58**, 26 (2020). https://doi.org/10.1109/mcom.001.2000297
36. S. Deng, H. Zhao, W. Fang, J. Yin, S. Dustdar, A.Y. Zomaya, IEEE Internet Things J. **7**, 7457 (2020). https://doi.org/10.1109/jiot.2020.2984887
37. D. Kolevski, K. Michael, I.E.E.E. Technol, Soc. Mag. **43**, 22 (2024). https://doi.org/10.1109/mts.2024.3372605
38. G. Loewenstein, D.A. Asch, K.G. Volpp, Health Aff. (Millwood) **32**, 1244 (2013). https://doi.org/10.1377/hlthaff.2012.1163
39. F. Sugahara, J. Pascual-Anaya, Y. Oisi, S. Kuraku, S.I. Aota, N. Adachi, W. Takagi, T. Hirai, N. Sato, Y. Murakami, S. Kuratani, Nature **531**, 97 (2016). https://doi.org/10.1038/nature16518
40. D. Ham, H. Park, S. Hwang, K. Kim, Nat. Electron. **4**, 635 (2021). https://doi.org/10.1038/s41928-021-00646-1
41. W. Gerstner, W.M. Kistler, *Spiking Neuron Models: Single Neurons* (Plasticity (Cambridge University Press, Populations, 2002)
42. B. Rajendran, A. Sebastian, M. Schmuker, N. Srinivasa, E. Eleftheriou, IEEE Signal Process. Mag. **36**, 97 (2019). https://doi.org/10.1109/msp.2019.2933719
43. A. Tavanaei, M. Ghodrati, S.R. Kheradpisheh, T. Masquelier, A. Maida, Neural Netw. **111**, 47 (2019). https://doi.org/10.1016/j.neunet.2018.12.002
44. S.P. Pati, T. Yajima, Jpn. J. Appl. Phys. (2008) **63**, 030801 (2024). https://doi.org/10.35848/1347-4065/ad297b
45. H. Inoue, H. Tamura, A. Kitoh, X. Chen, Z. Byambadorj, T. Yajima, Y. Hotta, T. Iizuka, G. Tanaka, I.H. Inoue, Adv. Mater.**n/a**(n/a), 2407326 (2024). https://doi.org/10.1002/adma.202407326
46. S. Ghose, A.G. Yaglikçi, R. Gupta, D. Lee, K. Kudrolli, W.X. Liu, H. Hassan, K.K. Chang, N. Chatterjee, A. Agrawal, M. O'Connor, O. Mutlu, Proc. ACM Meas. Anal. Comput. Syst. **2**, 1 (2018). https://doi.org/10.1145/3224419
47. H. Liu, A. Gegov, M. Cocea, Granular Comput. **1**, 259 (2016). https://doi.org/10.1007/s41066-016-0021-6
48. V.D. Blondel, J.N. Tsitsiklis, Automatica (Oxf.) **36**, 1249 (2000). https://doi.org/10.1016/s0005-1098(00)00050-9
49. J.R. Searle, Behav. Brain Sci. **3**, 417 (1980). https://doi.org/10.1017/s0140525x00005756
50. X. Chen, T. Yajima, I.H. Inoue, T. Iizuka, Jpn. J. Appl. Phys. **61**, SC1051 (2022). https://doi.org/10.35848/1347-4065/ac43e4
51. X. Chen, Z. Byambadorj, T. Yajima, H. Inoue, I.H. Inoue, T. Iizuka, Appl. Phys. Lett. **122**, 074102 (2023). https://doi.org/10.1063/5.0136627

52. S.P. Pati, Y. Geng, S. Hamasuna, K. Fujiwara, T. Iizuka, H. Inoue, I. Inoue, T. Yajima, Commun. Mater. **5**, 1 (2024). https://doi.org/10.1038/s43246-024-00621-1
53. H. Tamura, G. Tanaka, Neural Netw. **143**, 550 (2021). https://doi.org/10.1016/j.neunet.2021.06.031
54. I. Tsuda, Behav. Brain Sci. **24**, 793 (2001). https://doi.org/10.1017/s0140525x01000097
55. D.J. Amit, *Modeling brain function: The world of attractor neural networks* (Cambridge University Press, 1989). https://doi.org/10.1017/cbo9780511623257
56. X.J. Wang, Neuron **36**, 955 (2002). https://doi.org/10.1016/s0896-6273(02)01092-9
57. L.R. Squire, S.M. Zola, Proc. Natl. Acad. Sci. U. S. A. **93**, 13515 (1996). https://doi.org/10.1073/pnas.93.24.13515
58. H. Eichenbaum, Neuron **44**, 109 (2004). https://doi.org/10.1016/j.neuron.2004.08.028
59. P.W. Frankland, B. Bontempi, Nat. Rev. Neurosci. **6**, 119 (2005). https://doi.org/10.1038/nrn1607

Reservoir Computing Models for Slow Electronics

Gouhei Tanaka

Abstract With the penetration of AI technology into society, it has become necessary to consider using the right AI in the right place. Large-scale AI with extremely high-performance capabilities for cloud service consumes a lot of energy and requires a large amount of data for training. In contrast, edge AI placed at the edge of the Internet aims to adaptively process data at high speed using low-power computing devices, potentially mitigating increased data traffic for cloud AI. In this chapter, we first explain that reservoir computing is one of the promising lightweight machine learning frameworks for temporal pattern recognition. Next, we introduce reservoir computing models, including the echo state network and its variant models, with particular focus on their timescale conversion property for time series data. We demonstrate that extended echo state networks with broadly distributed time constants are able to better approximate multi-timescale nonlinear dynamical systems than conventional models. The demonstration suggests an importance of appropriate selection of machine learning models that match the timescale of biological and human-related data targeted by slow electronics.

1 Introduction

In the IoT (Internet of Things) society, many terminal devices and sensors in daily life are connected to the Internet. The amount of data obtained through measurement devices has increased rapidly with the development of sensing technology. It is desired that a rich variety of measured data are leveraged to improve the convenience of people's lives and the productivity of industries. High-end cloud AI can be constructed by aggregating a large amount of data in data centers and training large-scale deep learning models, which offer "**excellent inference**" services accessible to many

G. Tanaka (✉)
Graduate School of Engineering, Nagoya Institute of Technology, Nagoya, Japan
e-mail: gtanaka@nitech.ac.jp

International Research Center for Neurointelligence, Institutes for Advanced Study, The University of Tokyo, Tokyo, Japan

I. H. Inoue (ed.), *Slow Electronics with Reservoir Computing*,
https://doi.org/10.1007/978-981-96-8383-3_2

people. Cloud AI excels in scalability and accessibility, but requires a vast amount of data and computational resources, leading to concerns about increased energy consumption and communication cost [1]. In addition, ensuring privacy protection for confidential and personal data remains a challenge. To address these issues associated with data growth and aggregation, efficient data processing at the edge of the Internet has recently received considerable attention [2].

Edge AI aims to process sensor signals at high speed with low-power processors within or near terminal devices, smartwatches, and wearable devices. Unlike cloud AI, edge AI performs "**standard inference**" sufficient to achieve the objective while adaptively updating machine learning (ML) models in response to environmental changes. Data selection and transformation with edge AI can contribute to reducing the amount of data transmitted through communication networks and mitigating the increased data traffic attributed to cloud AI. Since computational resources and data amount are not plentiful at the edge, ML models available for edge AI are often limited in practice. Figure 1 illustrates the concept of "**the right AI in the right computer**", which means that the appropriate AI/ML model differs depending on the computer size. For example, fast-trainable compact ML models are better suited for application-specific microcomputers and edge AI devices than large-scale general-purpose ML models.

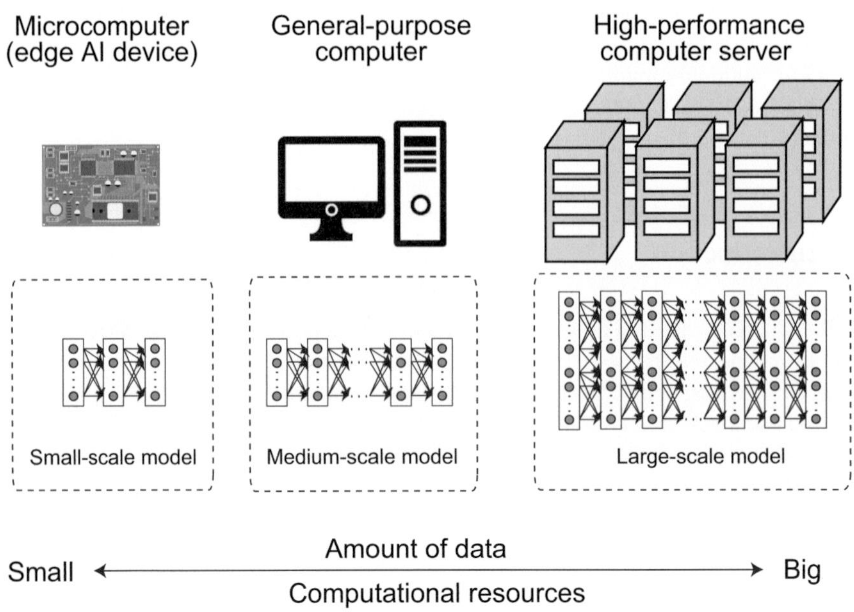

Fig. 1 Concept of "the right AI in the right computer," indicating the relationship between computer size and appropriate AI/ML model size. Appropriate models need to be selected considering the limitation on the amount of available data, computational resources, and computer power

In this chapter, we focus on reservoir computing which is one of the lightweight ML frameworks promising for edge AI [3–6]. Reservoir computing is an efficient training method for recurrent neural networks (RNNs) and also a computational principle for constructing AI hardware with various physical devices that are not limited to electronic circuits. The echo state network (ESN) has been the most widely used reservoir computing model applicable to a variety of time series data processing tasks [7]. In recent years, many advanced ESN models have been developed in terms of model architecture and dynamics to enhance computational performance and improve computational efficiency. ESNs are able to acquire generalization ability to approximate dynamical systems (or dynamic filters) by learning from training data, with training cost much lower than that of RNNs. The main topic of this chapter is the timescale conversion property of the ESN models, which are useful for handling time series data associated with multi-timescale dynamical systems. To demonstrate the effectiveness of this property, we apply ESN models to time series prediction with data generated by a phenomenological neuron model. The results show that reservoirs with rich time constant distributions in variant ESN models enhance the performance in time series prediction.

The remainder of this chapter is organized as follows. In Sect. 2, we clarify the positioning of the ESN in comparison with other neural network models. In Sect. 3, we introduce the standard ESN and several variant ESNs. In Sect. 4, we analyze time constant distributions of the ESN models. In Sect. 5, we demonstrate the results of numerical experiments on the time series prediction. In Sect. 6, we conclude this chapter.

2 Neural Networks for Time Series Pattern Recognition

With the development of sensory devices, the status of objects are becoming observable with monitoring devices in real time. For example, sensors mounted on smartwatches and wearable devices can measure biometric and human activity data over time. Downsized signal measurement instruments, such as those for medicine and healthcare, enable data logging in various environments. Recordings of time-varying signals are represented as time series data where current data have certain relation with past data. Under this circumstance, increasing attention has been paid to ML models for handling time series data. Such ML models typically have a mechanism to learn patterns in data variations over time, which characterize the dynamical system behind the observed time series data.

Figure 2 illustrates representative neural network models. The perceptron is the first computational model that utilizes artificial neurons [8]. The output of this model is given by a sum of weighted inputs. The weights are adjusted to yield desired outputs through a learning algorithm. The perceptron is categorized into a linear model and is not able to solve linearly inseparable problems. It is followed by the multilayer feedforward neural networks (MFNN) which can solve linearly inseparable problems using the backpropagation algorithm [9]. MFNNs with a large number of hidden

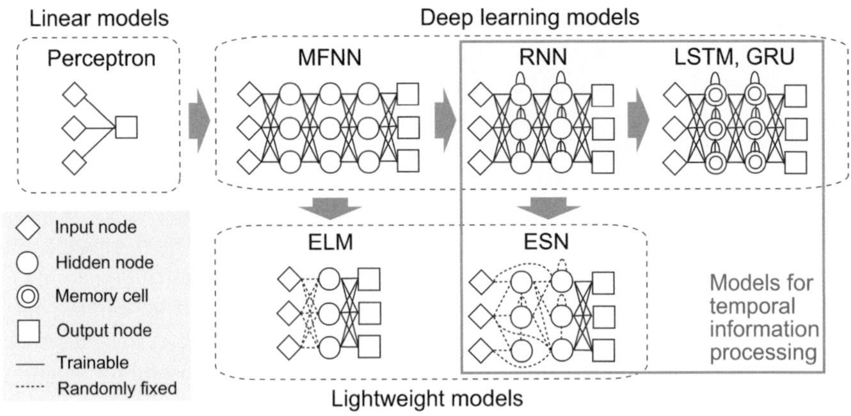

Fig. 2 Artificial neural network models. The models can be categorized into three types: linear models, deep learning models, and lightweight models. Linear models include the perceptron. Deep learning models include multilayer feedforward neural network (MFNN), recurrent neural networks (RNN), long short-term memory (LSTM), and gated recurrent unit (GRU). The lightweight models include extreme learning machine (ELM) and echo state network (ESN). Among these, RNN, LSTM, GRU, and ESN are suitable for temporal pattern recognition

layers are called deep learning models. The extreme learning machine (ELM) is a three-layer FNN where only the output connection weights are trained [10]. This training approach allows a reduction in training cost compared to the MFNN. These feedforward-type models are not suitable for treating time series data because of the absence of internal feedback connections.

RNNs have internal feedback connections in the hidden layer, which allow for past inputs to influence the present hidden states and enable pattern recognition for a sequence of inputs [11]. In other words, the architecture of this model offers a key mechanism to possess dynamic memory. The model is typically unrolled over time into a layered architecture and trained with the backpropagation through time (BPTT) algorithm. Therefore, the RNN is categorized into a deep learning model. All the connection weights in the model are optimized in the training phase. RNNs often face difficulty in training due to vanishing or exploding gradient problems and in handling long-term interdependency of data.

These limitations were mitigated by extended RNN models including the long short-term memory (LSTM) and the gated recurrent unit (GRU) [12, 13]. The LSTM introduces memory cells that have gating mechanisms to control the proportion of input memory and forgetting. The memory cells allow the model to capture long-term interdependency of data. The GRU simplifies the gating mechanism by adopting a different model architecture. These models improve the computational performance in temporal pattern recognition in exchange for increased training cost. The LSTM and GRU are also categorized into deep learning models.

ESNs simplify the training of RNNs by randomly setting the connection weights except for the output connection weights [3, 7, 14]. An ESN model can be regarded as a combination of a nonlinear transformation of inputs by a fixed recurrent network (called a *reservoir*) and a linear sum of weighted reservoir states (called a *readout*). The ESN and other relevant models based on the similar concept are generically known as reservoir computing models [15]. The readout weights are optimized with linear learners, leading to a considerable reduction in training cost compared to RNNs. The reservoir can generate a dynamical state that can store past input information (i.e. an echo state). In exchange for the low training cost, the performance of an ESN is normally not better than that of a fully trained RNN of the same size because trainable parameters are limited. The ESN and the ELM share the characteristic of having fixed random weights and restricted trainable weights, categorized into lightweight ML models.

In summary, among the neural network models mentioned above, the RNN, LSTM, GRU, and ESN can suitably process time series data for pattern recognition because of their mechanism to store past inputs in dynamic memory embedded in the states of recurrent neural networks. In particular, the ESN is a favorable option that meets the demand for edge AI handling time series data due to their computational efficiency.

3 Echo State Network and Its Variants

In this section, we introduce the standard ESN and variant ESN models with rich time constants. First, we introduce single-reservoir ESNs, including the original ESN in Sect. 3.1, the leaky-integrator ESN (LI-ESN) in Sect. 3.2, and the diverse-timescale ESN (DTS-ESN) in Sect. 3.3. Then, we present multi-reservoir models, including the grouped LI-ESN (G-LI-ESN) in Sect. 3.4 and the grouped DTS-ESN (G-DTS-ESN) in Sect. 3.5. The architectures of these models are illustrated in Fig. 3.

3.1 Echo State Network (ESN)

The ESN is a representative of reservoir computing models (Fig. 3a) [3, 7]. A standard ESN model consists of an input layer, a reservoir (or a hidden recurrent layer), and an output layer. Input data given to the input layer are nonlinearly transformed into a high-dimensional feature space in the reservoir. Then, a weighted sum of the reservoir states is taken to produce model output. The time evolution of the reservoir states and the linear summation in the readout are expressed, respectively, as follows:

$$\mathbf{x}_{n+1} = \mathbf{f}(W_{\text{in}}\mathbf{u}_{n+1} + W_{\text{rec}}\mathbf{x}_n + W_{\text{fb}}\mathbf{y}_n), \quad (1)$$

$$\mathbf{y}_{n+1} = W_{\text{out}}\mathbf{x}_{n+1}, \quad (2)$$

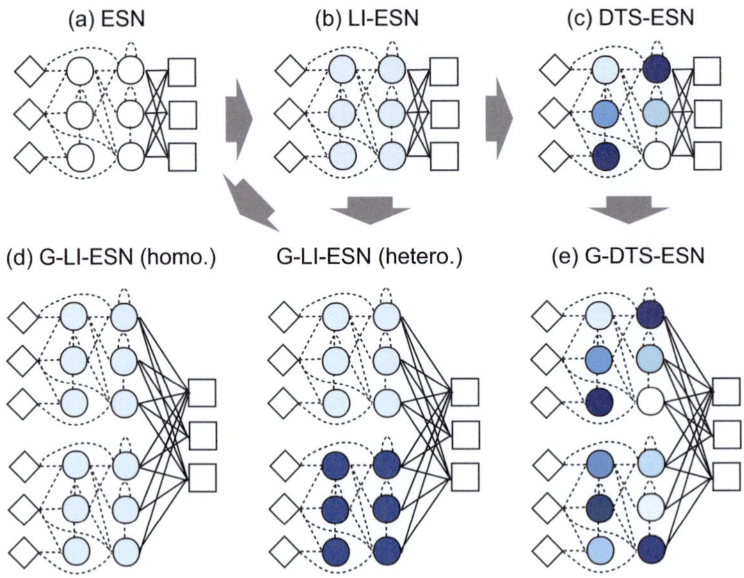

Fig. 3 Architectures of ESN models. Output feedback connections are omitted for clarity. **a** A standard ESN [7]. **b** A leaky-integrator ESN (LI-ESN) [16] **c** A diverse-timescale ESN (DTS-ESN) [17]. **d** A grouped LI-ESN (G-LI-ESN) [18] with homogeneous (left) and heterogeneous (right) reservoirs. **e** A grouped DTS-ESN (G-DTS-ESN)

where $\mathbf{u}_n \in \mathbb{R}^{N_u}$, $\mathbf{x}_n \in \mathbb{R}^{N_x}$, and $\mathbf{y}_n \in \mathbb{R}^{N_y}$ represent the input vector, the reservoir state vector, and the output vector at discrete time n, respectively. The function $\mathbf{f}$ represents an element-wise nonlinear activation function. The hyperbolic tangent function is normally used as the activation function. The matrices W_{in}, W_{rec}, W_{fb}, and W_{out} denote the weight matrices for the input-to-reservoir connections, the recurrent connections in the reservoir, the output-to-reservoir connections, and the output-to-reservoir connections, respectively. The output feedback connections are optional.

The architecture of the ESN is similar to that of the RNN. The ESN uses sparse recurrent networks in the reservoir, whereas the RNN assumes a fully connected recurrent network in the hidden layer. In the ESN, only W^{out} in Eq. (2) is optimized with a linear learner, while W^{in}, W^{rec}, and W^{fb} are randomly initialized and fixed. This setting is different from that of the RNN where all connection weight matrices are trained with the BPTT [11]. Therefore, in exchange for the reduction in the number of trainable parameters, the computational performance of ESN relies largely on the design of the reservoir.

An empirical recipe for hyperparameter settings in ESNs is found in the practical guide [19]. For example, the elements of W^{in} are randomly drawn from the uniform distribution with range $[-\gamma, \gamma]$ where γ is called the *input scaling factor*. When the connection *density* of the reservoir network is d, W^{rec} contains non-zero elements with a proportion of d. These non-zero values are randomly drawn from a uniform distribution and then scaled such that the *spectral radius* of W^{rec} equals ρ. The value of ρ should be set to ensure the echo state property [20] (or the fading memory property [21]). The elements of W^{fb} are also randomly drawn from the uniform distribution with range $[-\varphi, \varphi]$ where φ is called the *feedback scaling factor*.

The main prerequisites for a reservoir include nonlinearity, high dimensionality, and short-term memory as follows [3]:

- **Nonlinearity**: A nonlinear transformation of inputs in the reservoir is attributed to the operation of the nonlinear activation function $\mathbf{f}$. The input scaling factor γ, the spectral radius ρ, and the feedback scaling factor φ change the range of the input to the function and influence the strength of the nonlinearity.
- **High-dimensionality**: The dimension of the reservoir state vector is determined by the number of reservoir nodes, N_x, which is set at a sufficiently large number. To avoid redundancy in the reservoir states, the connection density d is set at a small value.
- **Short-term memory**: The internal feedback connections in the reservoir allow to generate an echo of inputs, from which we can reproduce recent past inputs. This property controlled by the spectral radius ρ is related to how much distant past information can be considered in time series data processing.

An ESN model is trained such that the model output vector matches the target output vector at each time step. By collecting the equations for all time steps in the training phase, the condition to be satisfied is written as follows:

$$D = W_{\mathrm{out}} X, \tag{3}$$

where $X = [\mathbf{x}_1, \mathbf{x}_2, \ldots, \mathbf{x}_T] \in \mathbb{R}^{N_x \times T}$ is the reservoir state vector collection matrix, $D = [\mathbf{d}_1, \mathbf{d}_2, \ldots, \mathbf{d}_T] \in \mathbb{R}^{N_y \times T}$ is the teacher vector collection matrix, and T is the length of training data. In batch training, an approximate solution of Eq. (3) is obtained by minimizing the squared error plus the regularization term as follows:

$$E = \|D - W_{\mathrm{out}} X\|_2^2 + \frac{\beta}{2} \|W_{\mathrm{out}}\|_F, \tag{4}$$

where β is the *regularization factor* and $\|\cdot\|_F$ denotes the Frobenius norm. The output matrix optimized via ridge regression is written as follows:

$$\hat{W}_{\mathrm{out}} = D X^\top (X X^\top + \beta I)^{-1}. \tag{5}$$

The time complexity of this computation is $O(N_x^2 T)$ when $N_x \gg N_u$ and $N_x \gg N_y$. There are other methods to optimize the readout part, such as recursive least squares and FORCE learning [22–24].

3.2 Leaky Integrator ESN (LI-ESN)

In the standard ESN (Sect. 3.1), the speed of the time evolution of the reservoir state is almost determined by the rate of changes in the input sequence. However, in a general temporal task, input and desired time series can have different timescales. In such a case, it is useful for the model to have a mechanism that controls the input response speed of the reservoir.

The LI-ESN is an extension of the standard ESN, that uses leaky-integrator units in the reservoir (Fig. 3b) [16]. To cope with the timescale gap between inputs and outputs, the leaking property is introduced. The time evolution of the LI-ESN is represented as follows:

$$\mathbf{x}_{n+1} = (1-\alpha)\mathbf{x}_n + \alpha\mathbf{f}(W_{\text{in}}\mathbf{u}_{n+1} + W_{\text{rec}}\mathbf{x}_n + W_{\text{fb}}\mathbf{y}_n), \tag{6}$$

$$\mathbf{y}_{n+1} = W_{\text{out}}\mathbf{x}_{n+1}, \tag{7}$$

where $\alpha \in (0, 1]$ is the leaking rate that adjusts the balance between maintaining and updating the previous reservoir state. The leaking rate α influences how the reservoir responds quickly to input changes. When $\alpha = 1$, the LI-ESN is reduced to the standard ESN in Eqs. (1)–(2). The smaller the leaking rate, the slower the reservoir state changes. The leaking rate is one of the hyperparameters that significantly affect computational performance of the LI-ESN, which needs to be appropriately tuned [25].

3.3 Diverse-Timescale ESN (DTS-ESN)

The reservoir nodes in the LI-ESN (Sect. 3.2) have the identical leaking rate. This setting leads to homogeneous input responses in the reservoir nodes. Therefore, the LI-ESN is not able to work well when desired outputs have multiple timescales. To address multi-timescale systems, a reservoir with a variety of input responses is desirable.

The DTS-ESN is an extension of the LI-ESN with distributed decay time constants of the reservoir (Fig. 3c) [17]. The update equation for the reservoir state vector in the DTS-ESN is expressed as follows:

$$\mathbf{x}_{n+1} = (I - A)\mathbf{x}_n + A\mathbf{f}(W_{\text{in}}\mathbf{u}_{n+1} + W_{\text{rec}}\mathbf{x}_n + W_{\text{fb}}\mathbf{y}_n), \tag{8}$$

$$\mathbf{y}_{n+1} = W_{\text{out}}\mathbf{x}_{n+1}, \tag{9}$$

where $A = \text{diag}(\alpha_1, \ldots, \alpha_{N_x})$ is a diagonal matrix of leaking rates $\alpha_i (i = 1, \ldots, N_x)$. The leaking rates are drawn from a certain probability distribution in the range of $[\alpha_{\min}, \alpha_{\max}]$. In Ref. [17], it is assumed that α_i follows a reciprocal (or log-uniform) distribution because it is empirically known that the performance of the LI-ESN is affected by changes in the leaking rates on a logarithmic scale. The DTS-ESN can output time series with diverse timescales by setting a sufficiently wide range for the leaking rates. By doing this, fine tuning of leaking rates is not necessary. The DTS-ESN has flexibility to accommodate a variety of desired outputs and especially has an advantage in approximating multi-timescale systems.

3.4 Grouped LI-ESN (G-LI-ESN)

In general, increasing the number of reservoir nodes improves the comoputational performance of ESNs. This is because the number of trainable parameters increases, allowing to express various nonlinear dynamical systems. However, it is empirically known that performance improvement plateaus as the number of nodes increases. To address this issue, multi-reservoir ESNs consisting of multiple reservoirs have been investigated [18]. The constituent reservoirs can be connected in series (DeepESN), in parallel (GroupedESN), or in a more complicated way [26–28]. A computationally efficient multi-reservoir model, such as the multi-step-learning ESN [29], has also been developed.

The G-LI-ESN consists of multiple reservoirs arranged in parallel as shown in Fig. 3d. Two types of models with identical reservoirs and heterogeneous reservoirs are depicted in the left and the right, respectively. The original inputs are given to all the reservoirs. The state vectors of the multiple reservoirs are concatenated at each time step, and the concatenated vectors are collected to form a reservoir state collection matrix. The model equations are written as follows:

$$\mathbf{x}_{n+1}^{(r)} = (1 - \alpha^{(r)})\mathbf{x}_n^{(r)} + \alpha^{(r)}\mathbf{f}(W_{\text{in}}^{(r)}\mathbf{u}_{n+1} + W_{\text{rec}}^{(r)}\mathbf{x}_n^{(r)} + W_{\text{fb}}^{(r)}\mathbf{y}_n), \ r = 1, \ldots, N_r, \tag{10}$$

$$\mathbf{y}_{n+1} = W_{\text{out}}[\mathbf{x}_{n+1}^{(1)}; \ldots; \mathbf{x}_{n+1}^{(N_r)}], \tag{11}$$

where N_r denotes the number of reservoirs and r ($r = 1, \ldots, N_r$) represents the index of the reservoir. The number of nodes in the rth reservoir is denoted by $N_x^{(r)}$, satisfying $\sum_r N_x^{(r)} = N_x$. Equation (10) can be viewed as an N_x-dimensional equation with respect to $\mathbf{x}_n = [\mathbf{x}_{n+1}^{(1)}; \ldots; \mathbf{x}_{n+1}^{(N_r)}]$. The hyperparameters and the connection weight matrices can be shared in part or in whole by the reservoirs to prevent excessive large number of hyperparameters.

3.5 Grouped Diverse-Timescale ESN (G-DTS-ESN)

The G-DTS-ESN is an extension of the G-LI-ESN with distributed leaking rates as shown in Fig. 3e. The model equations are written as follows:

$$\mathbf{x}_{n+1}^{(r)} = (I - A^{(r)})\mathbf{x}_n^{(r)} + A^{(r)}\mathbf{f}(W_{\text{in}}^{(r)}\mathbf{u}_{n+1} + W_{\text{rec}}^{(r)}\mathbf{x}_n^{(r)} + W_{\text{fb}}^{(r)}\mathbf{y}_n), \; r = 1, \ldots, N_r, \tag{12}$$

$$\mathbf{y}_{n+1} = W_{\text{out}}[\mathbf{x}_{n+1}^{(1)}; \cdots ; \mathbf{x}_{n+1}^{(N_r)}], \tag{13}$$

where $A^{(r)}$ denotes the diagonal matrix of leaking rates of the rth reservoir, drawn from a probability distribution in the range of $[\alpha_{\text{min}}, \alpha_{\text{max}}]$. Equation (12) can be viewed as an N_x-dimensional equation with respect to $\mathbf{x}_n = [\mathbf{x}_{n+1}^{(1)}; \ldots ; \mathbf{x}_{n+1}^{(N_r)}]$. This model is considered a special case of the DTS-ESN.

4 Time Constant Analysis

4.1 Linearization

In this section, we analyze time constant distributions of the reservoirs in the variant ESN models introduced in Sect. 3. The state update equations for the reservoir dynamics in the ESN models are regarded as nonlinear dynamical systems. They are represented as N_x-dimensional discrete-time dynamical systems, where the state of a reservoir node evolves under the interaction with the states of other nodes. The time evolution of reservoir states in a short time period is almost governed by the linear term in the righthand side of the reservoir state update equations. The linearized equations are described in general as follows:

$$\mathbf{x}_{n+1} = J\mathbf{x}_n, \tag{14}$$

where J is the Jacobian matrix. All the models in Sect. 3 are written in this form after linearization. Using a sparse random matrix W with spectral radius 1, the recurrent connection weights can be written as $W_{\text{rec}} = \rho W$. The Jacobian matrices for the four models in Sect. 3 can be specifically obtained as follows:

- **LI-ESN** (Sect. 3.2)

$$J_{\text{LI}} = (1 - \alpha)I + \rho\alpha W. \tag{15}$$

- **DTS-ESN** (Sect. 3.3)

$$J_{\text{DTS}} = (1 - A)I + \rho A W. \tag{16}$$

- **G-LI-ESN** (Sect. 3.4)

$$J_{\text{G-LI}} = \begin{bmatrix} (1-\alpha^{(1)}) + \rho^{(1)}\alpha^{(1)} W & \dots & O \\ O & \ddots & O \\ O & \dots & (1-\alpha^{(N_r)}) + \rho^{(N_r)}\alpha^{(N_r)} W \end{bmatrix}. \quad (17)$$

- **G-DTS-ESN** (Sect. 3.5)

$$J_{\text{G-DTS}} = \begin{bmatrix} (I-A^{(1)}) + \rho^{(1)}A^{(1)} W & \dots & O \\ O & \ddots & O \\ & \dots & (I-A^{(N_r)}) + \rho^{(N_r)}A^{(N_r)} W \end{bmatrix}. \quad (18)$$

4.2 *Time Constant Distributions*

The Jacobian matrix is related to the exponential decay rate of the linearized system in Eq. (14). When the Jacobian matrix J is diagonalizable, we obtain the following equation:

$$P^{-1} J P = \Lambda, \quad (19)$$

where $\Lambda = \text{diag}(\lambda_1, \dots, \lambda_{N_x})$ is a diagonal matrix of the eigenvalues of J and P is a certain regular matrix. Equations (14) and (19) yield the following equation:

$$\mathbf{z}_{n+1} = \Lambda \mathbf{z}_n, \quad (20)$$

where $\mathbf{z}_n = P\mathbf{x}_n$. The solution of the ith equation is represented as follows:

$$\begin{aligned} z_n^{(i)} &= \lambda_i^n z_0^{(i)} \\ &= |\lambda_i|^n (\cos(n\theta_i) + j\sin(n\theta_i)) z_0^{(i)}, \quad \text{for } i = 1, \dots, N_x, \end{aligned} \quad (21)$$

where j denotes the imaginary unit. The absolute value of $z_n^{(i)}$ increases exponentially with time n if $|\lambda_i| > 1$ and decays exponentially if $|\lambda_i| < 1$. By assuming an exponential decay of $z_n^{(i)}$ and introducing a decay time constant τ_i, we have the following equation:

$$e^{-n/\tau_i} = |\lambda_i|^n \quad \text{for } i = 1, \dots, N_x. \quad (22)$$

Therefore, the decay time constant is approximately represented as follows [17]:

$$\tau_i = -\frac{1}{\ln|\lambda_i|}, \quad \text{for } i = 1, \ldots, N_x. \tag{23}$$

The set of decay time constants characterizes the input responsiveness of the reservoir.

4.3 Numerical Analysis

Once the hyperparameters of the reservoirs are determined, we can numerically compute time constants of reservoir dynamics for each variant ESN model using Eq. (23) and Eqs. (15)–(18). The eivenvalues of W, $\lambda_i(W)$ for $i = 1, \ldots, N_x$, are uniformly distributed over the unit disk on the complex plane. For the time constant analysis, we set the total number of reservoir nodes as $N_x = 2000$, the density of the recurrent connections in the reservoir as $d = 0.1$, and the spectral radius as $\rho = 1$ in all the models. Under this setting, we obtained eigenvalue distributions, the ranges of time constant distributions with different leaking rate settings, and the histogram of time constants for the four variant ESNs as shown in Fig. 4.

Figure 4a shows the result for LI-ESNs. The eigenvalues of J_{LI} are uniformly distributed on a disk with radius $\rho\alpha$ and center $(1-\alpha, 0)$ in the complex plane. With a decrease of α value from 1, the area over which the eigenvalues are distributed shrinks. In the limit of $\alpha \to 0$, all the eigenvalues converge to (1,0) along the real axis. When $|\lambda_i| \sim 1$ and $\mathrm{Im}(\lambda) \sim 0$, we have $\ln|\lambda_i| \sim |\lambda_i| - 1$ and $|\lambda_i| \sim 1 - \mathrm{Re}(\lambda)$. Hence, the time constants in Eq. (23) can be approximated as follows:

$$\tau_i = \frac{1}{-\ln|\lambda_i|} \sim \frac{1}{1 - \mathrm{Re}(\lambda_i)}, \tag{24}$$

which is consistent with the formula in Ref. [30]. In the LI-ESN, the time constant distribution is concentrated on the value determined by the single leaking rate α and is limited to a relatively narrow range because the eigenvalues of J_{LI} are distributed in a narrow region.

Figure 4b shows the result for DTS-ESNs. We assume that leaking rates follow a reciprocal (or log-uniform) distribution in the range of $[\alpha_{\min}, \alpha_{\max}]$ with $\alpha_{\max} = 1$. In other words, $\log\alpha$ follows a uniform distribution. This assumption is motivated by empirical knowledge that α affects computational performance on the logarithmic scale. We aim to cover a wide range of input response times. Since the leaking rates are distributed in this model, the eigenvalue distribution keeps a relatively large range. As $\alpha_{\min}$ decreases, the range of the leaking rate distribution increases. Accordingly, a time constant distribution becomes broader. The rich time constants are the remarkable characteristics of the DTS-ESN.

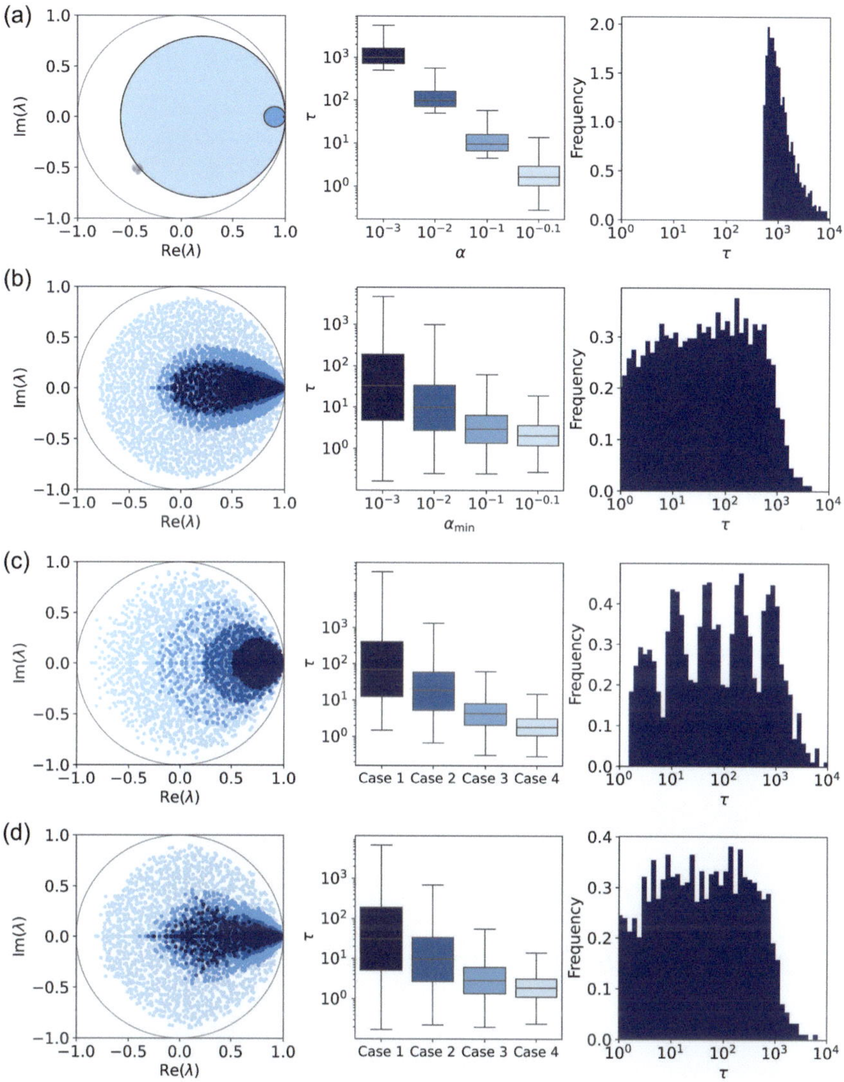

Fig. 4 Characteristics of the linearized systems of the ESN models. The left panels show distributions of eigenvalue of the Jacobian matrix. The center panels show the ranges of the decay constant distributions with different settings of leaking rates. The right panels demonstrate the histograms of time constants that correspond to the leftmost case in the center panels. **a** LI-ESN with a single leaking rate α. **b** DTS-ESN with distributed leaking rates in $[\alpha_{\min}, 1]$. **c** G-LI-ESN with five reservoirs with different leaking rates. **d** G-DTS-ESN with five reservoirs with distributed leaking rates

Figure 4c shows the result for G-LI-ESNs. We set the number of reservoirs at $N_r = 5$. Each reservoir has $N_x^{(r)} = 400$ nodes for $r = 1, \ldots, N_r$. The leaking rates were set as follows:

$$\left(\log \alpha^{(1)}, \ldots, \log \alpha^{(N_r)}\right) = \left(\log \alpha, \frac{\log \alpha}{N_r}, \ldots, \frac{(N_r - 1)\log \alpha}{N_r}\right), \tag{25}$$

where $\log \alpha = -3$ for Case 1, $\log \alpha = -2$ for Case 2, $\log \alpha = -1$ for Case 3, and $\log \alpha = -0.1$ for Case 4. The eigenvalues of $J_{\mathrm{G-LI}}$ are widely distributed compared to those of J_{LI}. The time constant range becomes larger as $\log \alpha$ decreases. In Case 1, the time constant distribution has multiple peaks, indicating that the time constant distributions of the individual reservoirs are narrow but their superposition brings about a relatively broad distribution.

Figure 4d shows the result for G-DTS-ESNs. We set the number of reservoirs at $N_r = 5$. Each reservoir has $N_x^{(r)} = 400$ nodes for $r = 1, \ldots, N_r$. The maximum leaking rates are set as $\log \alpha_{\max}^{(r)} = 0$ for all r. The minimum leaking rates are shared by the reservoirs as follows:

$$\log \alpha_{\min}^{(1)} = \log \alpha_{\min}^{(2)} = \cdots = \log \alpha_{\min}^{(N_r)} = \log \alpha_{\min}, \tag{26}$$

where $\log \alpha_{\min} = -3$ for Case 1, $\log \alpha_{\min} = -2$ for Case 2, $\log \alpha_{\min} = -1$ for Case 3, and $\log \alpha_{\min} = -0.1$ for Case 4. The eigenvalue distribution of $J_{\mathrm{G-DTS}}$ is close to that of J_{DTS}. In Case 1, the range of the time constant distribution is as large as that of DTS-ESNs and G-LI-ESNs. The time constant distribution of G-DTS-ESN is similar to that of DTS-ESN but less uniform.

5 Time Series Prediction

5.1 Multi-timescale Dynamical System

We evaluate computational ability of the four variant ESN models in a time series prediction task related to multi-timescale systems. Multi-timescale dynamical systems are ubiquitous in real-world systems where dynamical components with different timescales are interacting with each other. In nonlinear dynamical systems theory, such multi-timescale systems can be understood with the fast-slow analysis [31]. One of the challenges concerned with multi-timescale systems in biology and medicine is to integrate ML and multiscale modeling techniques [32].

As an example, we generate time series data by simulating the Hindmarsh-Rose phenomenological neuron model [33]. The model equation is described as follows:

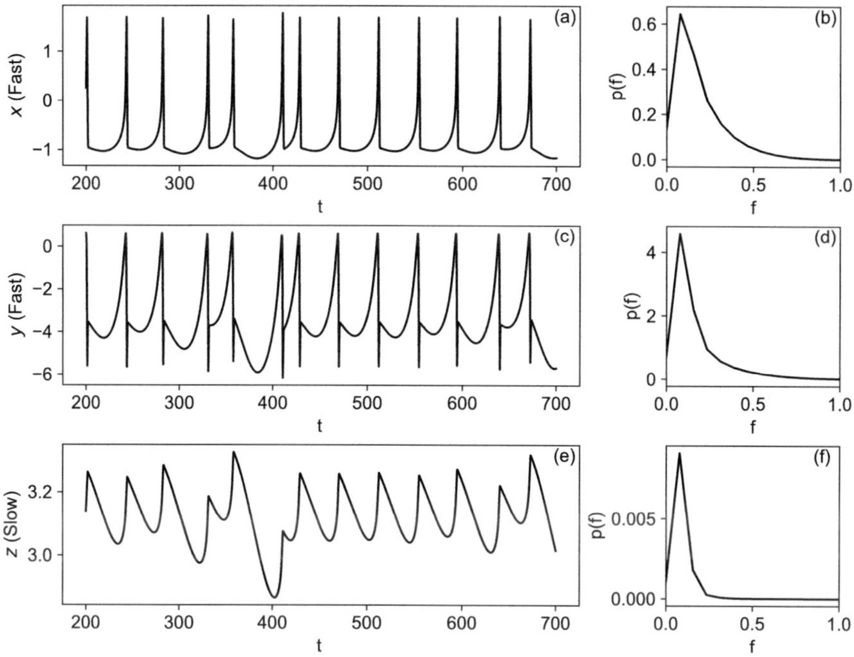

Fig. 5 An example of multi-timescale (fast-slow) dynamical systems. The time courses of the Hindmarsh-Rose neuron model and the corresponding power spectra are shown. **a–b** The fast variable x. **c–d** The fast variable y. **e–f** The slow variable z

$$
\begin{aligned}
\frac{\mathrm{d}x}{\mathrm{d}t} &= y - x^3 + bx^2 - z + I, \\
\frac{\mathrm{d}y}{\mathrm{d}t} &= 1 - 5x^2 - y, \\
\frac{\mathrm{d}z}{\mathrm{d}t} &= \epsilon(s(x - x_0) - z),
\end{aligned}
\tag{27}
$$

where $b = 3.05$, $I = 3$, $\epsilon = 0.01$, $s = 4$, and $x_0 = -1.6$. It is known that the model exhibits chaotic dynamics under this parameter condition. This model is a fast-slow dynamical system where x and y are the fast variables and z is the slow variable. The time courses and power spectral densities are exemplified in Fig. 5.

5.2 Time Series Prediction

Typical ML tasks with time series data are mainly categorized into time series generation, reconstruction, prediction (forecasting), classification, and anomaly detection. To investigate the timescale conversion properties of the four models in Sect. 3, we performed a time series prediction task.

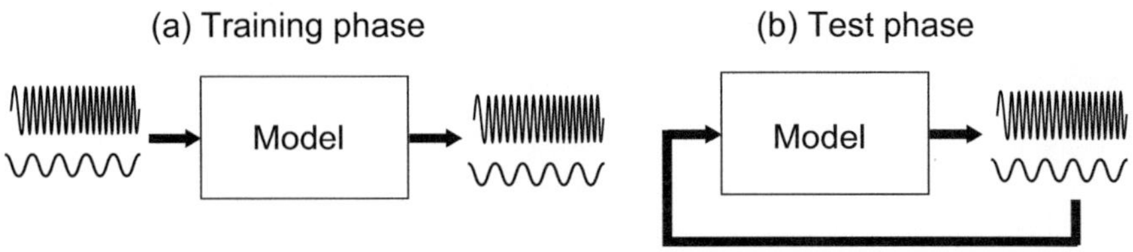

Fig. 6 Time series prediction task. **a** A model is trained to predict the one-step-ahead state of the fast-slow dynamical system. **b** With a given initial condition, the trained model can generate time series autonomously with the closed feedback loop

Time series data of total length 10200 were generated by numerically integrating the HR model with time step 0.05. After discarding the initial data of length 200 in the transient period, we divided the rest of data equally into five segments. We divided each segment evenly into training and test data.

Figure 6 illustrates the task setting. In the training phase (Fig. 6a), a model is trained to produce the one-step-ahead value of the input time series. In the test phase (Fig. 6b), the trained model with a given initial condition produces output sequences autonomously due to the closed loop. The initial condition was taken from the test data.

Due to the chaotic behavior of the target system, a long-term prediction is not possible. This is because the initial error between the model output and the target output increases exponentially. The prediction performance is evaluated with the Valid Time, which indicates the time interval until the error between prediction and target exceeds a threshold [29, 34]. For simplicity, we evaluate the error regarding the variable z and set the threshold value to 0.4.

We performed numerical experiments using variant ESN models, including LI-ESN, DTS-ESN, G-LI-ESN, and G-DTS-ESN. The total number of nodes was fixed at 480 in all models. In multi-reservoir models (i.e. G-LI-ESN and G-DTS-ESN), the number of reservoirs was varied from one to six. As the number of reservoirs increases, the number of reservoir nodes in each reservoir decreases. The hyperparameter values of each model were adjusted using the Bayesian optimization method [35, 36]. The optimization process consists of an exploration phase and a subsequent exploitation phase. The number of exploration steps was set at 16 and that of exploitation steps was also at 16. Table 1 shows the hyperparameters of the variant ESN models. For the multi-reservoir models, we assumed that the values of the input scaling factor γ, the spectral radius ρ, the feedback scaling factor φ, and the regularization factor β were common to all the reservoirs.

Table 1 Hyperparameters of the ESN variant models. γ is the input scaling parameter, ρ is the spectral radius, φ is the feedback scaling factor, α is the leaking rate, β is the regularization parameter. The number of reservoirs is denoted by N_r in multi-reservoir ESNs

Model	# Hyperparameters	Hyperparameters
LI-ESN	5	$\gamma, \rho, \varphi, \log\beta, \log\alpha$
DTS-ESN	4	$\gamma, \rho, \varphi, \log\beta$
G-LI-ESN	$4 + N_r$	$\gamma, \rho, \varphi, \log\beta, \log\alpha^{(r)}$ $(r = 1, \ldots, N_r)$
G-DTS-ESN	4	$\gamma, \rho, \varphi, \log\beta$

5.3 Results

We demonstrate the time courses generated by a trained G-DTS-ESN composed of four reservoirs ($N_r = 4$) in Fig. 7. In this case, the Valid Time is around 704, corresponding to the shaded region where the predictions (dark colored curves) are in good agreement with the target time series (light colored curves).

For multi-reservoir models, we have an option to appropriately set the number of reservoirs. Figure 8 shows the effect of the number of reservoirs (N_r) in the G-LI-ESN and the G-DTS-ESN on prediction performance. The five panels correspond to the five segments used for the experiments. We can see that the performance of the G-LI-ESN tends to almost monotonically degrade with an increase in the number of reservoirs in all the panels. Therefore, the best result is obtained for the single-reservoir model with $N_r = 1$, which is equivalent to the LI-ESN. On the other hand, the performance of the G-DTS-ESN is best at $N_r = 4$ for Segments 1, 3,

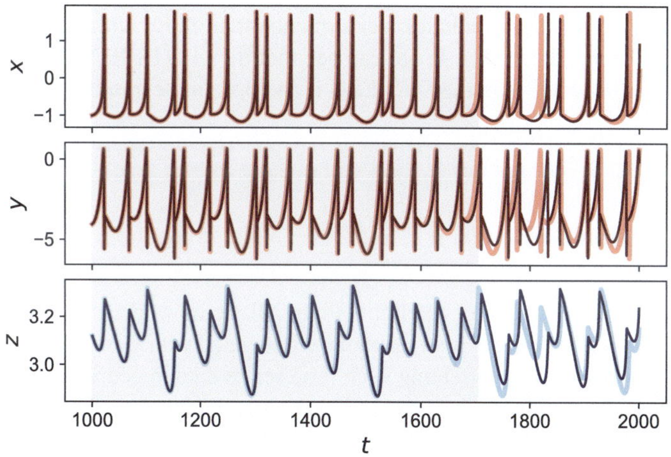

Fig. 7 An example of time series prediction with a G-DTS-ESN consisting of four reservoirs ($N_r = 4$). The dark colored curves indicate the model predictions, and the light colored curves indicate the target time series

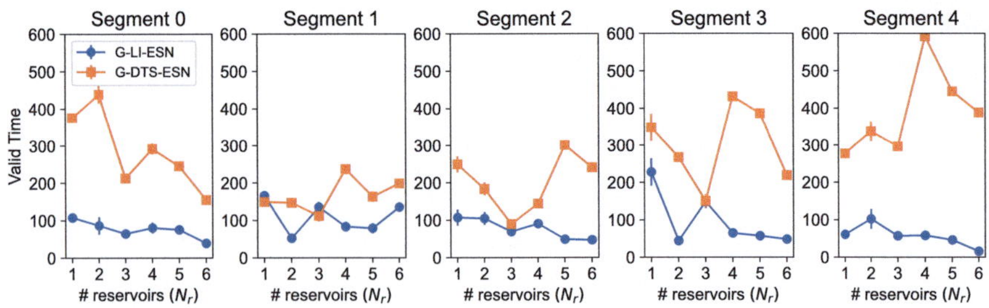

Fig. 8 Effect of the number of reservoirs (N_r) in multi-reservoir models on the Valid Time. Five panels correspond to the five segments. The blue circles and orange squares indicate the Valid Times for G-LI-ESNs and G-DTS-ESNs, respectively

and 4. This means that the incorporation of diverse time constants (or distributed leaking rates) into the multi-reservoir models can improve prediction performance. The results suggest that the structural heterogeneity of reservoirs and the wide time constant distribution of each reservoir provide a substantial synergy in generating a rich variety of responses to inputs, enhancing the ability to approximate multi-timescale dynamical systems.

Figure 9 shows a comparison of the four models. It contains five panels that correspond to the five segment data. Each panel shows how the Valid Times obtained in the exploitation phase of the Bayesian optimization are distributed for the four models. The boxplot indicates statistics of the Valid Time distribution, including the maximum, average, and minimum values. The Valid Time distribution differs depending on the used segment data, because the segments have different irregularity. In comparison between the four models, we can see that DTS-ESNs and G-DTS-ESNs outperfom the two other models. This means that the models with distributed leaking rates yield better prediction performance than those with identical leaking rates. From this result and Fig. 8, we can see that the G-DTS-ESN can yield better prediction performance than the DTS-ESN if an appropriate number of reservoirs is chosen.

6 Conclusions

Reservoir computing is one of the lightweight ML frameworks that is a promising basis for edge AI. Real-world time series data associated with slow electronics have a variety of timescales. When treating such time series data in neuromorphic systems [37], the problem of timescales arises [38]. In this chapter, we introduced variant ESN models that are able to fill a timescale gap between input and output sequences. The numerical results suggest that reservoirs with rich time constants are suitable for treating time series data obtained from multi-timescale systems.

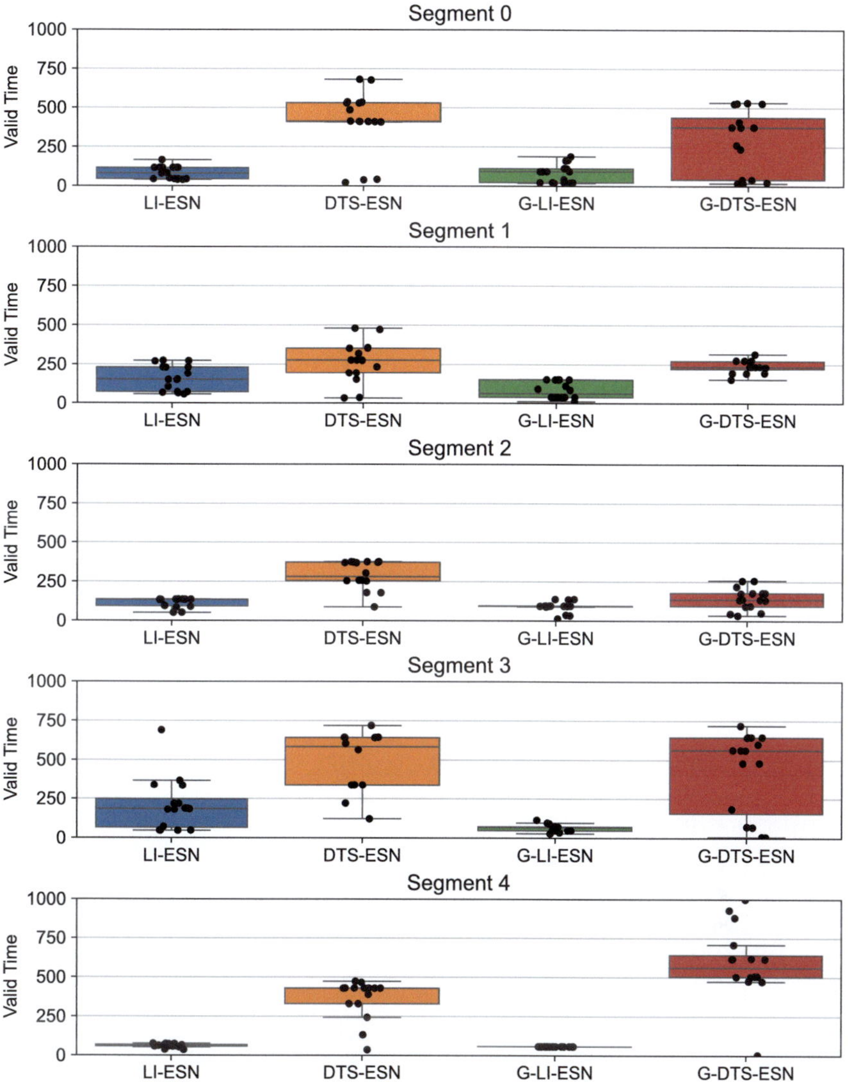

Fig. 9 Comparison of ESN models using Valid Time in the long-term prediction task. The models include LI-ESN, DTS-ESN, G-LI-ESN, and G-DTS-ESN. The five panels correspond to five segments used for the experiments. The boxplots indicate the statistics of the distribution of 16 Valid Times obtained in the exploitation phase of the Bayesian optimization. The circles indicate the individual Valid Times

Competing Interests

The author has no conflicts of interest to declare that are relevant to the content of this chapter.

Acknowledgements We thank Keigo Iwashima for fruitful discussion. This research was partly supported by JST CREST (JPMJCR19K2, JPMJCR24R2), JSPS KAKENHI (JP23K28154), and JST Moonshot R&D (JPMJMS2021).

References

1. A. Katal, S. Dahiya, T. Choudhury, Cluster Comput. **26**(3), 1845 (2023)
2. Y. Shi, K. Yang, T. Jiang, J. Zhang, K.B. Letaief, IEEE Commun. Surv. Tutor. **22**(4), 2167 (2020)
3. M. Lukoševičius, H. Jaeger, Comput. Sci. Rev. **3**(3), 127 (2009)
4. G. Tanaka, T. Yamane, J.B. Héroux, R. Nakane, N. Kanazawa, S. Takeda, H. Numata, D. Nakano, A. Hirose, Neural Netw. **115**, 100 (2019)
5. G. Tanaka, C. Gallicchio, A. Micheli, J.P. Ortega, A. Hirose, IEEE Trans. Neural Netw. Learn. Syst. **33**(6), 2571 (2022)
6. K. Nakajima, I. Fischer, *Reservoir computing* (Springer, 2021)
7. H. Jaeger, Bonn, Germany: German National Research Center for Information Technology GMD Technical Report **148**(34), 13 (2001)
8. F. Rosenblatt, Psychol. Rev. **65**(6), 386 (1958)
9. D.E. Rumelhart, G.E. Hinton, R.J. Williams, Nature **323**(6088), 533 (1986)
10. G.B. Huang, Q.Y. Zhu, C.K. Siew, Neurocomputing **70**(1–3), 489 (2006)
11. P.J. Werbos, Proc. IEEE **78**(10), 1550 (1990)
12. S. Hochreiter, *Neural Computation* (MIT-Press, 1997)
13. J. Chung, C. Gulcehre, K. Cho, Y. Bengio (2014). arXiv:1412.3555
14. H. Jaeger, H. Haas, Science **304**(5667), 78 (2004)
15. D. Verstraeten, B. Schrauwen, M. d'Haene, D. Stroobandt, Neural Netw. **20**(3), 391 (2007)
16. H. Jaeger, M. Lukoševičius, D. Popovici, U. Siewert, Neural Netw. **20**(3), 335 (2007)
17. G. Tanaka, T. Matsumori, H. Yoshida, K. Aihara, Phys. Rev. Res. **4**(3), L032014 (2022)
18. C. Gallicchio, A. Micheli, L. Pedrelli, Neurocomputing **268**, 87 (2017)
19. M. Lukoševičius, in *Neural Networks: Tricks of the Trade: Second Edition* (Springer, 2012), pp. 659–686
20. I.B. Yildiz, H. Jaeger, S.J. Kiebel, Neural Netw. **35**, 1 (2012)
21. W. Maass, T. Natschläger, H. Markram, Neural Comput. **14**(11), 2531 (2002)
22. H. Jaeger, Advances in Neural Information Processing Systems **15** (2002)
23. D. Sussillo, L.F. Abbott, Neuron **63**(4), 544 (2009)
24. H. Tamura, G. Tanaka, Neural Netw. **143**, 550 (2021)
25. M. Lukoševičius, H. Jaeger, B. Schrauwen, KI-Künstliche Intelligenz **26**, 365 (2012)
26. Z. Li, G. Tanaka, Neurocomputing **467**, 115 (2022)
27. Z. Li, Y. Liu, G. Tanaka, Appl. Soft Comput. **135**, 110021 (2023)
28. Z. Li, K. Fujiwara, G. Tanaka, in *2024 International Joint Conference on Neural Networks (IJCNN)* (IEEE, 2024), pp. 1–9
29. T. Akiyama, G. Tanaka, IEEE Access **10**, 28535 (2022)
30. L. Manneschi, M.O. Ellis, G. Gigante, A.C. Lin, P. Del Giudice, E. Vasilaki, Front. Appl. Math. Stat. **6**, 616658 (2021)
31. R. Bertram, J.E. Rubin, Math. Biosci. **287**, 105 (2017)
32. M. Alber, A. Buganza Tepole, W.R. Cannon, S. De, S. Dura-Bernal, K. Garikipati, G. Karniadakis, W.W. Lytton, P. Perdikaris, L. Petzold, et al., NPJ Digital Med. **2**(1), 115 (2019)

33. J.L. Hindmarsh, R. Rose, Proc. Roy. Soc. Lond. Ser. B. Biol. Sci. **221**(1222), 87 (1984)
34. J. Pathak, A. Wikner, R. Fussell, S. Chandra, B.R. Hunt, M. Girvan, E. Ott, Chaos: An Interdiscip. J. Nonlinear Sci. **28**(4) (2018)
35. F. Nogueira, et al. (2014)
36. N. Stander, K.J. Craig, Eng. Comput. **19**(4), 431 (2002)
37. D.V. Christensen, R. Dittmann, B. Linares-Barranco, A. Sebastian, M. Le Gallo, A. Redaelli, S. Slesazeck, T. Mikolajick, S. Spiga, S. Menzel et al., Neuromorphic Comput. Eng. **2**(2), 022501 (2022)
38. H. Jaeger, D. Doorakkers, C. Lawrence, G. Indiveri (2021). arXiv:2102.10648

Fabricating Elements of Slow Electronics with Functional Materials

Hisashi Inoue and Takeaki Yajima

Abstract This chapter introduces the concept of slow electronics, focusing on the key principles behind slow electronic devices and the underlying technologies. It explores the slow dynamics of ions, particularly oxygen ions and protons, and demonstrates how these phenomena in solids can be harnessed to create fundamental slow electronic devices.

1 Why Slow Electronics Devices?

Slow electronics relies on slow electronic devices as its fundamental components. Although this statement may seem self-evident, its implications are not widely recognized. Consider the conventional metal-oxide-semiconductor (MOS) field-effect transistor (FET), commonly used in modern consumer electronics. The basic structure of a MOSFET is illustrated in Fig. 1(a). It consists of gate, drain, and source electrodes. The current flowing from the drain to the source (I_D) is controlled by the voltage applied to the gate electrode. For example, when a voltage V_0 is applied at time $t = 0$, I_D switches on almost instantaneously. The switching speed of conventional MOSFETs can reach approximately 1 ns, enabling operation at GHz clock frequencies. Moreover, as the device size decreases, the switching speed generally increases due to reduced capacitance.

Now, consider a FET with a slower switching speed, as shown in Fig. 1(b). When a gate voltage is applied at $t = 0$, I_D gradually increases over time. After a period corresponding to the device's characteristic time scale (τ), I_D reaches saturation. If the time dependence of I_D remains consistent after the voltage is applied, it is possible to determine when the voltage was applied by observing the value of I_D.

H. Inoue (✉)
National Institute of Advanced Industrial Science and Technology (AIST), Tsukuba, Japan
e-mail: hisashi.inoue@aist.go.jp

T. Yajima
Department of Electrical Engineering and Computer Science, Kyushu University, Fukuoka, Japan
e-mail: yajima@ed.kyushu-u.ac.jp

I. H. Inoue (ed.), *Slow Electronics with Reservoir Computing*,
https://doi.org/10.1007/978-981-96-8383-3_3

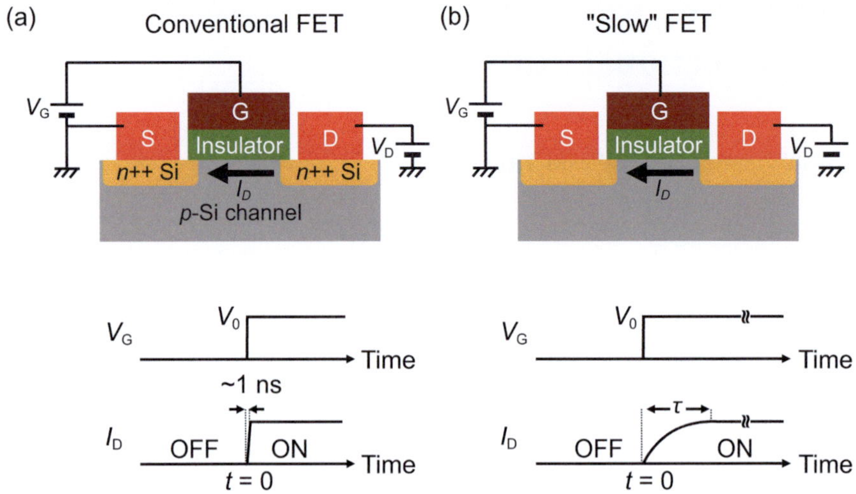

Fig. 1 Comparison of (**a**) a conventional FET with (**b**) a "slow" FET when a stepwise voltage is applied to the gate (G) terminal. S and D denote the source and drain terminals, respectively

However, once $t > \tau$, this information is lost because I_D becomes independent of time. This phenomenon is known as short-term memory and plays a crucial role in slow electronics. Notably, short-term memory is absent in fast FETs, where I_D remains constant immediately after $t = 0$.

One might assume that an infinitely long characteristic time scale ($\tau \to \infty$) is ideal for slow electronic devices, as this would enable the device to retain information indefinitely, effectively functioning as non-volatile memory [1–4]. However, non-volatility is not always beneficial for information processing in slow electronics. In Fig. 1(b), when $\tau \to \infty$, I_D changes minimally during a given time interval, making it impossible to determine when the voltage was applied based solely on I_D. Consequently, non-volatility (or long-term memory) is distinct from short-term memory and cannot be used interchangeably. The optimal value of τ depends on the design of the processing unit, including its resolution, but it generally corresponds to the characteristic time scale of the signals processed by slow electronics.

Slow electronics aims to enable real-time information processing with minimal energy consumption, significantly lower than that required by the conventional Von Neumann architecture. In traditional computing systems, information is processed using fast MOSFETs and stored in non-volatile memories, as illustrated in Fig. 2(a). In this architecture, the central processing unit (CPU), composed of approximately 10^{10} fast MOSFETs, performs complex signal processing, while input and output data are stored in non-volatile memory devices such as hard disk drives (HDDs),

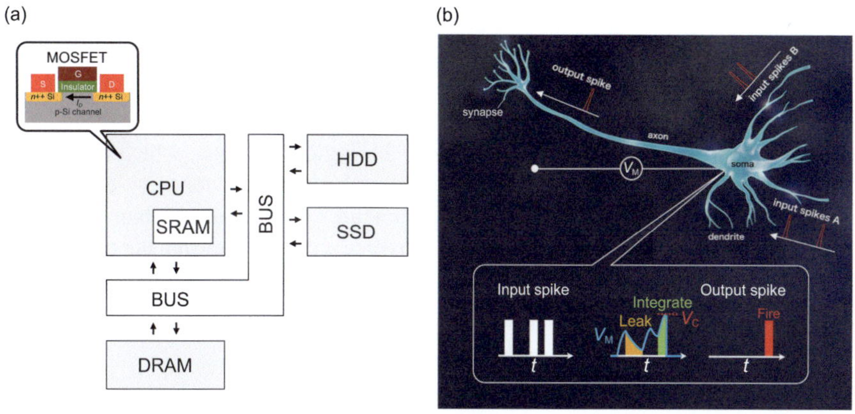

Fig. 2 Schematic representation of (**a**) the Von Neumann architecture and (**b**) a biological neuron. Inset: Schematic of the leaky-integrate-and-fire behavior. Adopted from [5]

solid-state drives (SSDs), and static random-access memories (SRAMs).[1] The CPU and memory units are physically separated and connected via data buses, forming the well-known Von Neumann architecture.

While conventional computers can emulate the behavior of slow electronic devices with short-term memory by periodically updating the value of I_D, this approach is energy-inefficient. Emulation requires retrieving the current value of I_D from memory, processing the update in the CPU, and writing the new value back to memory. In contrast, slow electronic devices store information within the device itself, eliminating the need for data retrieval and storage operations. As a result, slow electronic devices have the potential to achieve ultra-low energy consumption, a key advantage in slow electronics.

2 Biological Neuron

Slow electronics is designed to process human-interactive tasks. The time scales of human-oriented signals, such as motion and voice patterns, typically range from 1 ms to 1 s. To build an effective slow electronics processing system, a large number of slow electronic device elements are required. For optimal energy efficiency, each element must also consume minimal power. Additionally, the physical size of each element must be compact to allow for high-density integration.

In conventional central processing units (CPUs), approximately 10^{10} transistors serve as the fundamental processing elements. If the total power consumption of a

[1] Dynamic random-access memory (DRAM) is also used in conventional computers. Although SRAM and DRAM are classified as volatile memory, the term "volatile" is used differently here, as SRAM and DRAM lose information when power is turned off.

CPU is around 100 W, then the power consumption of each transistor must be less than 10 nW to maintain energy efficiency.

One of the most energy-efficient processing systems operating on a 1 ms to 1 s time scale is the biological brain, which consumes only about 20 W of power [6]. The brain is composed of a complex network of biological neurons, which are among the most energy-efficient elementary devices with short-term memory capabilities on the order of 1 ms to 1 s [7]. Each biological neuron consumes less than 100 pW of power. As a result, despite comprising approximately 10^{11} neurons, the total power consumption of the brain remains remarkably low.

3 Leaky-Integrate-and-Fire Model of Neurons

The response of biological neurons to external stimuli was studied by A. L. Hodgkin and A. F. Huxley in 1952 [8]. Their Hodgkin-Huxley model, based on experiments with the giant axon of a squid, remains the most fundamental model of neuronal behavior. Without delving into the details of the Hodgkin-Huxley model,[2] this section presents a simplified version known as the leaky-integrate-and-fire model.

Biological neurons maintain an electric potential called the membrane potential (V_M), measured relative to the extracellular environment. This potential arises from the concentration of various ions (e.g., Na^+, K^+, and Cl^-) inside the neuron. The flow of these ions in and out of the cell is regulated by ion channels embedded in the cell membrane. V_M increases when positive ions enter or negative ions leave the cell, and decreases when negative ions enter or positive ions leave. The rate at which V_M changes depends on the ionic conductance of the channels and the effective capacitance of the cell membrane. Consequently, the modulation of V_M by external stimuli occurs over a characteristic time scale, typically ranging from 1 ms to 1 s, which aligns with the time scale of human-oriented signals.

Figure 2(b) illustrates the typical behavior of V_M when a biological neuron receives a voltage pulse. During the pulse, V_M gradually increases, whereas it gradually returns to its equilibrium potential once the pulse ends. This gradual increase and decrease in V_M is known as leaky integration, which serves as a short-term memory mechanism for the neuron. The time scale of leaky integration varies depending on the neuron type, typically ranging from 1 ms to 100 ms [10–12]. When V_M reaches a critical threshold value (V_C), the neuron "fires" and emits a voltage pulse, known as a spike, which signals neighboring neurons of the firing event.

Inspired by the exceptional energy efficiency of biological neural networks and neurons, artificial neural networks that mimic the leaky-integrate-and-fire behavior offer a promising approach for building energy-efficient slow electronics [13]. The remainder of this chapter explores how these artificial neurons can be realized by leveraging ion migration in solid-state devices.

[2] For a comprehensive explanation of the Hodgkin-Huxley model, refer to [9].

4 Ionic Drift Diffusion

To implement an artificial neuron, the leaky-integration function must first be realized. The time scale of this leaky integration must be sufficiently long (1 ms to 1 s) to effectively process human-oriented signals targeted by slow electronics. However, achieving energy-efficient leaky integration within a compact device is challenging. Various designs and implementations of leaky-integration devices have been proposed [14–28], some of which are summarized in Table 1.

Ferroelectric devices utilize the polarization switching and relaxation properties of ferroelectric materials [29–35]. Atomic switch devices leverage the electrochemical formation of metallic filaments within an insulating material, modulating current flow in response to external voltages [36–44]. Meanwhile, Mott and phase-transition devices exploit the sensitivity of electronic and thermodynamic structural phase transitions to external stimuli, such as electric fields and temperature changes [45–49].

The most conventional approach for implementing leaky integration is to use a low-pass filter circuit consisting of a capacitor (C) and a resistor (R), as shown in Fig. 3(a). When a stepwise voltage change occurs at $t = 0$, from $V = 0$ to V_0, the voltage across the capacitor (V_{cap}) follows the exponential function:

$$V_{cap} = V_0 \left[1 - \exp\left(-\frac{t}{RC} \right) \right]. \tag{1}$$

In this equation, the product RC determines the characteristic time scale of the leaky integration. To achieve a time scale of 1 s, R and C must be as large as $R = 1\ \text{M}\Omega$ and $C = 1\ \mu\text{F}$, respectively. However, with conventional microprocessing technology, a 1 μF capacitor occupies approximately 4 mm^2 of space, making it impractical to integrate many artificial neurons within a compact device. Additionally, the electrical current flowing through the large resistor results in Joule heating during each charge and discharge cycle of the capacitor, leading to energy inefficiency. Therefore, alternative approaches are necessary for implementing long time-scale leaky integration.

Ionic drift-diffusion presents a promising solution to this challenge.[3] Taking a rechargeable battery as an example (Fig. 3(b)), charged ions migrate toward the electrode in response to an external voltage. This movement, driven by electrostatic force, is known as drift. When the external voltage is removed, and no electrostatic potential is present across the electrodes, the ions redistribute to achieve a uniform concentration. This movement, driven by the concentration gradient, is known as diffusion.

Ions generally migrate much more slowly than electrons, enabling long time-scale responses without the need for large capacitors. Additionally, ionic drift-diffusion is a reversible process that, in principle, consumes minimal energy, enhancing overall energy efficiency.

[3] Biological neurons utilize ionic drift-diffusion to achieve short-term memory, reflecting an evolutionary optimization for energy efficiency.

Table 1 Summary of various leaky-integration devices reported in the literature. Adopted from [5]

Mechanism	Material	Type^{*1}	Timescale	Pulse width^{*2}	Pulse height^{*2}	Pulse period^{*2}	Typical current w/o V_D^{*2}	Energy consumption w/o V_D^{*3}	Power consumption w/o V_D^{*4}	V_D^{*2}	Typical current w/ V_D^{*2}	Energy consumption w/ V_D^{*5}	Power consumption w/ V_D^{*6}	Area (channel $L \times W$)	Endurance	Reference
Ferroelectric FET	TaN/$Hf_{0.5}Zr_{0.5}O_2$/TaN	3T	0.4 s	5 ms	1.6 V	40 ms	–	–	–	1.8 V	–	–	–	–	–	[29]
Anti-ferroelectric FET	TiN/$Hf_xZr_{1-x}O_2$/TiN	3T	0.8 ms	1 µs	1.7 V	2 µs	–	–	–	0.3 V	50 nA	37 fJ/spike^{*7}	1.85 nW	10 µm^2	$> 10^{12}$	[34]
Floating gate FET	Graphene/hBN/MoS_2	2T or 3T	1.3 s	100 ms	4 V	4 s	1 nA	400 pJ/spike	100 pW	–	–	–	–	400 µm^2	$> 10^4$	[25]
Protonic resistance change	Pt/WO_3/Pd	2T	49 s	5 s	0.45 V^{*8}	50 s	20 µA	10 µJ/spike^{*9}	200 nW^{*10}	–	–	–	–	0.03 mm^2	–	[26]
Atomic diffusion and redox	Pt/Ag/Ag:SiO_2/Ag/Pt	2T	300 µs ~ 30 ms	50 µs	1 V	100 µs	20 µA	1 nJ/spike	10 µW	–	–	–	–	25 µm^2	$> 10^6$	[31][38]
Mott transistion	$GaTa_4Se_8$ ($T = 74$ K)	2T	518 µs	20 µs	40 V	100 µs	400 µA	320 nJ/spike	3.2 mW	–	–	–	–	–	–	[47]
Phase change	TiO_2/VO_2/Au	2T	43 µs	10 µs	10 V	20 µs	1 mA	100 nJ/spike	5 mW	–	–	–	–	–	–	[49]
Ion migration	Ti/TiO_x/Pt	2T	40 ms	10 µs	4 V	100 µs	500 nA	20 pJ/spike	200 nW	–	–	–	–	25 µm^2	$> 5 \times 10^6$	[24]
Ion migration	SiO_2/WO_x/Pd/Au	2T	50 ms	500 µs	1.4 V	5 ms	1 µA	700 pJ/spike	140 nW	–	–	–	–	0.25 µm^2	–	[44]
Electrolysis	$SrTiO_3$/12CaO·7Al_2O_3	3T	21 s	65 s	10 V	240 s	200 pA	130 nJ/spike	542 pW	2 V	100 µA	13 mJ/spike	54 µW	0.16 mm^2	–	[27]
Ion migration	ITO/$CH_3NH_3PbI_3$/Au	2T	25 ms	5 ms	3 V	25 ms	30 µA	450 nJ/spike	18 µW	–	–	–	–	0.01 mm^2	> 4000	[28]
Mott transition	W/TiN/NbO_2/TiN with integrated capacitor	2T	1 µs	50 ns	2.05 V	1 µs	1.25 mA	130 pJ/spike	130 µW	–	–	–	–	0.003 µm^2	–	[42]
Drift diffusion	$SrTiO_3$/ParyleneC/HfO_x	3T	1.07 s	2 ms	5 V	0.1 s	30 pA	300 fJ/spike	3 pW	0.5 V	50 nA	50 pJ/spike	500 pW	16 µm^2	$> 10^8$	This work [5]

^{*1} 2T and 3T stand for two-terminal and three-terminal devices, respectively
^{*2} Typical value experimentally demonstrated in the refence and used for the energy and power estimation in this table
^{*3} Energy consumption w/o V_D = Pulse width × Pulse height × Typical current w/o V_D unless otherwise indicated
^{*4} Power consumption w/o V_D = Energy consumption w/o V_D / Pulse period unless otherwise indicated
^{*5} Energy consumption w/ V_D = Pulse width × V_D × Typical current w/ V_D unless otherwise indicated
^{*6} Power consumption w/ V_D = Energy consumption w/ V_D/ Pulse period unless otherwise indicated
^{*7} Value reported in the reference
^{*8} 0.5 V at idle state and 0.05 V at pulse
^{*9} Energy consumption w/o V_D = Power consumption w/o V_D × Pulse period
^{*10} Value reported in the reference

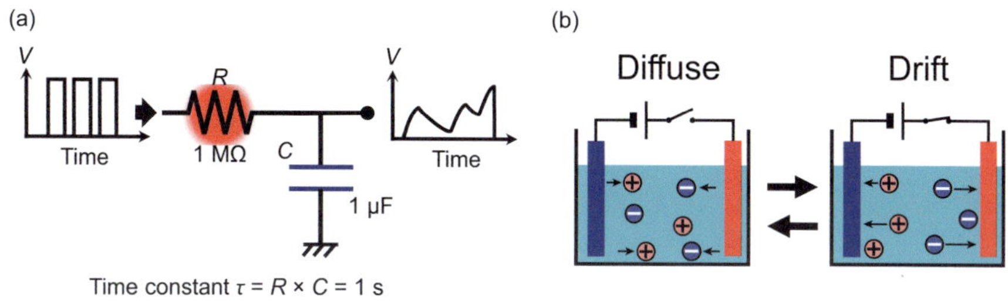

Fig. 3 (a) Low-pass filter implementation of leaky integration. (b) Schematic illustration of ionic drift-diffusion

In the following sections, we present two examples of ionic drift-diffusion devices designed for slow electronics.

5 Oxygen Vacancy Drift-Diffusion Field-Effect Transistor

5.1 Strontium Titanate Material

To select an appropriate material for ionic drift-diffusion devices, the material must be capable of accommodating ions, exhibit moderate ionic conductivity, and allow precise control over ionic distribution and transport. Additionally, it is crucial to devise a method for modulating electrical currents based on the ionic distribution and transport.

Strontium titanate ($SrTiO_3$) is a promising candidate material. It is an oxide semiconductor with a band gap of 3.2 eV and is often referred to as a "fake diamond" due to its optical refractive index, which closely resembles that of diamond [50, 51]. $SrTiO_3$ possesses a perovskite crystal structure, as shown in Fig. 4(a). In this structure, a titanium atom is surrounded by an oxygen octahedron within a cubic lattice of strontium atoms.

In realistic $SrTiO_3$ crystals, structural imperfections, including oxygen vacancies, are present at a typical density of approximately 10^{15} cm^{-3} [52]. Oxygen vacancies can be described by the chemical equation:

$$\mathrm{SrTiO_3(s)} \rightleftarrows [\mathrm{SrTiO_2}]^{2+}\,(\mathrm{s}) + \frac{1}{2}\mathrm{O_2(g)} + 2e', \tag{2}$$

or equivalently using Kröger-Vink notation:[4]

[4] $O_O^\times$ denotes an oxygen atom occupying an oxygen site, while $V_O^{\cdot\cdot 2+}$ represents an oxygen vacancy carrying a $+2e$ charge.

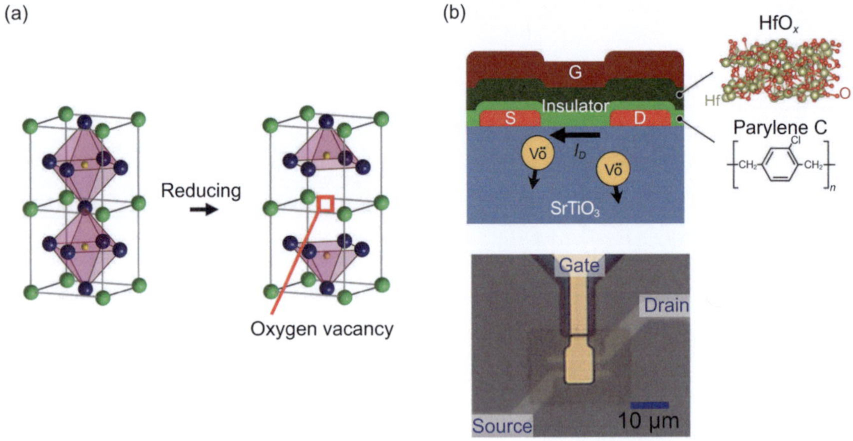

Fig. 4 (**a**) Schematic illustration of the $SrTiO_3$ crystal structure and oxygen vacancy generation by reduction. (**b**) Schematic structure of an $SrTiO_3$ FET and an optical microscope image of the fabricated device. Adopted from [5]

$$O_O^{\times} \rightleftarrows V_O^{\cdot\cdot} + \frac{1}{2}O_2(g) + 2e'. \tag{3}$$

An oxygen vacancy acts as a positively charged ion carrying a charge of $+2e$, where e is the elementary charge. Without electron doping, $SrTiO_3$ is an insulator at room temperature. However, oxygen-deficient $SrTiO_3$, typically produced by high-temperature annealing in a vacuum, exhibits metallic electrical properties due to electron doping.

Remarkably, oxygen vacancies in $SrTiO_3$ can migrate by exchanging positions with neighboring oxygen atoms within the crystal lattice or across its surfaces. Since these vacancies are positively charged ions, they undergo drift-diffusion under the influence of an external electric field. $SrTiO_3$ is particularly advantageous for controlling the drift-diffusion of oxygen vacancies because of its high relative permittivity of approximately 300. As discussed in Sect. 5.3, this high permittivity allows externally applied electric fields to extend over a large spatial range, enabling modulation of oxygen vacancy distribution across a wide area within the device. Since oxygen vacancies donate electrons to the conduction band of $SrTiO_3$, the material's electrical conductivity can be effectively modulated by controlling the distribution of these vacancies.

Using $SrTiO_3$ as a transistor channel, a field-effect transistor (FET) has been developed, as illustrated in Fig. 4(b) [5, 53–55]. The device is fabricated using conventional photolithography techniques and material deposition on a commercially available single-crystal $SrTiO_3$ substrate. The double-gate insulator layer, comprising Parylene C polymer and amorphous HfO_x, effectively minimizes electrical leakage and regulates oxygen exchange through the layer.

This device operates as an accumulation-mode FET, where a positive voltage applied to the gate electrode accumulates electrons in the conduction band of $SrTiO_3$, thereby increasing the electrical current flowing through the drain and source electrodes. Unlike the two-terminal devices reported elsewhere (see Table 1), the three-terminal configuration provides greater control over the electrical properties of the device. Specifically, this configuration includes a separate electrode for controlling channel conductivity, independent of the electrodes used to measure the resulting conductivity changes.

This separation allows the modulation of conductivity without direct injection of electrical current, thereby minimizing energy dissipation due to Joule heating. This characteristic is particularly advantageous for constructing neural networks, as it prevents the injection of parasitic currents into successive neurons during excitation, enhancing overall energy efficiency and signal integrity.

5.2 *Leaky-Integration Operation*

Figure 5(a) shows the response of the $SrTiO_3$ transistor to a few gate voltage pulses (V_G). Unlike conventional FETs, where the drain-to-source current (I_D) rises immediately upon activation, the I_D of the $SrTiO_3$ transistor gradually increases over several milliseconds after turning on. The transistor demonstrates an on/off ratio exceeding 10^6 with gate leakage below 30 pA, indicating excellent transistor performance.

To further characterize the dynamic behavior, the transistor was tested with a 80 Hz gate voltage pulses. As shown in Fig. 5(b), I_D switches on and off in response to each pulse, with the on-state current gradually increasing after each pulse. This behavior closely resembles the leaky-integration dynamics of membrane potentials

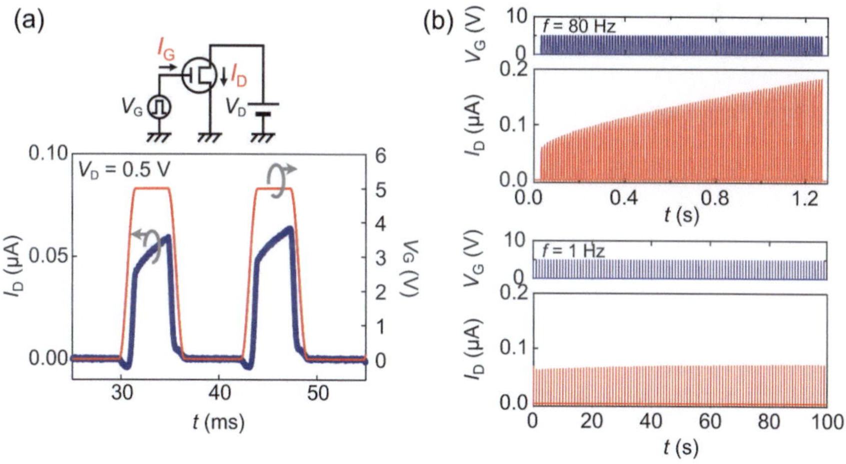

Fig. 5 (**a**) Current response of the $SrTiO_3$ FET to pulsed V_G excitations. (**b**) Leaky-integration behavior of the $SrTiO_3$ FET for a pulse train with varying excitation frequencies. Adopted from [5]

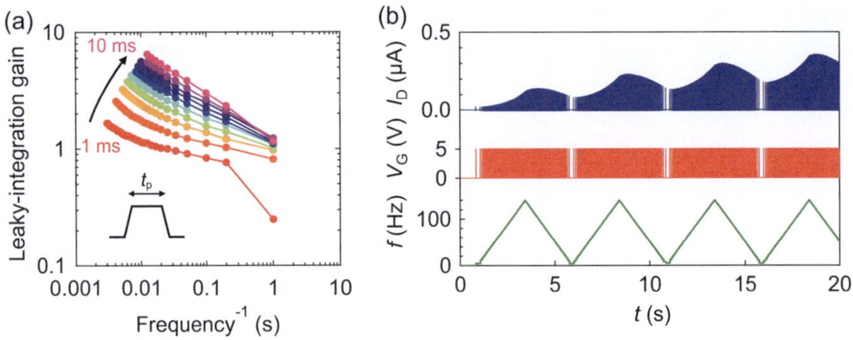

Fig. 6 (**a**) Frequency dependence of the leaky-integration gain for different pulse widths t_p. (**b**) I_D response to a continuously-frequency-modulated V_G pulse train. Adopted from [5]

in biological neurons. Notably, when the pulse frequency is reduced 1 Hz, the on-state current shows minimal increase, demonstrating the short-term memory effect discussed in Sect. 1.[5]

Figure 6(a) shows the leaky-integration gain as a function of the input pulse frequency f. The leaky-integration gain is defined as the ratio between the on-state current during the first gate pulse and that during the 100th gate pulse. The gain monotonically decreases with decreasing frequency and approaches unity around $f = 1$ Hz. For a given frequency, the gain increases monotonically with longer pulse widths. This behavior is consistent with the leaky-integration characteristic of biological neurons, except that I_D immediately returns to its equilibrium value once the gate voltage is removed, rather than gradually decaying.

The frequency-dependent leaky-integration gain is more clearly illustrated when the pulse frequency is continuously varied, as shown in Fig. 6(b). At higher frequencies, I_D increases more rapidly, while at lower frequencies, I_D decreases more quickly. These results indicate that the gain is influenced not only by the total duration of gate voltage application but also by the interval between gate voltage pulses, confirming a short-term memory effect with a characteristic time scale of approximately 1 s.

The time-scale of the leaky-integration behavior is determined more precisely by re-plotting the pulse response from Fig. 5(b) using the gate voltage (V_G) on the x-axis and the drain-to-source current (I_D) on a logarithmic y-axis. As seen from the pulse-by-pulse response curves in Fig. 7(a), the $I_D - V_G$ curves gradually shift toward lower voltages from the first to the 100th pulse, resulting in the observed leaky-integration effect.

To quantify this behavior, the threshold voltage V_{th} is defined as the value of V_G at which I_D crosses 10 nA, and it is plotted as a function of the pulse number N in Fig. 7(b). The dependence of V_{th} shows asymptotic behavior and fits well to a double exponential function:

[5] The transistor "forgets" the arrival of the previous pulse after its characteristic time scale.

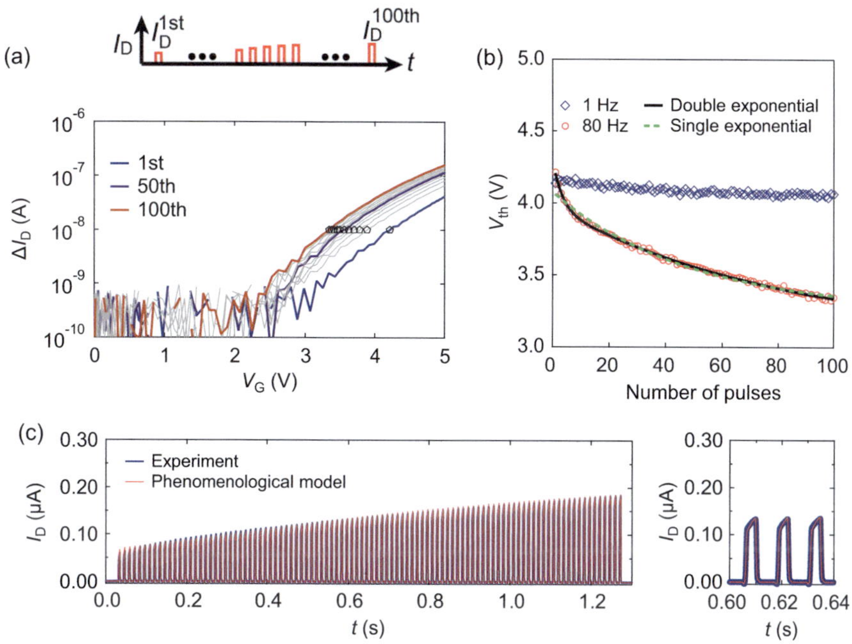

Fig. 7 Time-scale analysis of the leaky-integration. (**a**) Pulse-by-pulse $I_D - V_G$ curves obtained from the $f = 80$ Hz data shown in Fig. 5(b). (**b**) Double exponential shift of V_{th} during the leaky-integration. Fit results to the single and double exponential functions (Eq. (4)) are shown. (**c**) Comparison of the $f = 80$ Hz data shown in Fig. 5(b) and the phenomenological model Eqs. (5)-(6). Adopted from [5]

$$V_{th}(N) = V_0' + A_1 \exp(-\gamma_1 N) + A_2 \exp(-\gamma_2 N), \tag{4}$$

where A_1 and A_2 are the contributions from each exponential component, and the time constants are given by $\tau_1 \equiv 1/\gamma_1 f$ and $\tau_2 \equiv 1/\gamma_2 f$. Fitting the experimental data yields $\tau_1 = 984$ ms and $\tau_2 = 42$ ms, demonstrating long time-scale leaky-integration.

Based on the double exponential shift of V_{th}, a quantitative model for leaky-integration in the $SrTiO_3$ FET can be formulated. Since the shift of V_{th} depends on the history of the gate voltage, its time dependence is expressed as:

$$V_{th}(t) = V_0' - \int_{\infty}^{t} V_G(t') \left[\omega_1 \exp\left(-\frac{t-t'}{\tau_1'}\right) + \omega_2 \exp\left(-\frac{t-t'}{\tau_2'}\right) \right] dt', \tag{5}$$

where V_0' is the initial threshold voltage before gate voltage application, ω_1 and ω_2 are the relative contributions of each exponential, and τ_1' and τ_2' are the time constants. Here, an exponential dependence of V_{th} on time is assumed.

The drain-to-source current I_D is expressed using the FET model in both the sub-threshold and linear regions as [56]:

$$I_D(t) = -C\frac{dV_G(t)}{dt} + \begin{cases} I_0 \exp\left(\frac{V_G(t)-V_{th}(t)}{V_{ss}}\right) & (V_G(t) \le V_{th}(t)) \\ I_0\left[1 + z\left(V_G(t) - V_{th}(t)\right)\right] & (V_G(t) > V_{th}(t)), \end{cases} \tag{6}$$

where C is the effective gate capacitance, and I_0, V_{ss}, and z are phenomenological parameters. The first term describes the displacement current, which accounts for the dip in I_D just before the FET turns on and the hump observed just after it turns off, as seen in Fig. 5(a).

Using the time constants $\tau_1' = 1.07$ s and $\tau_2' = 5.0$ ms, the calculated I_D curve from Eqs. (5) and (6) reproduces the experimental data with high accuracy, as shown in Fig. 7(c).

5.3 Mechanism of Leaky Integration

This subsection examines the physical mechanism behind the exponential time dependence of the transistor characteristics observed in the experiment. As discussed in Sect. 5.1, $SrTiO_3$ contains a certain amount of oxygen vacancies (Fig. 8(a)). These vacancies act as positively charged ions that migrate within the crystal in response

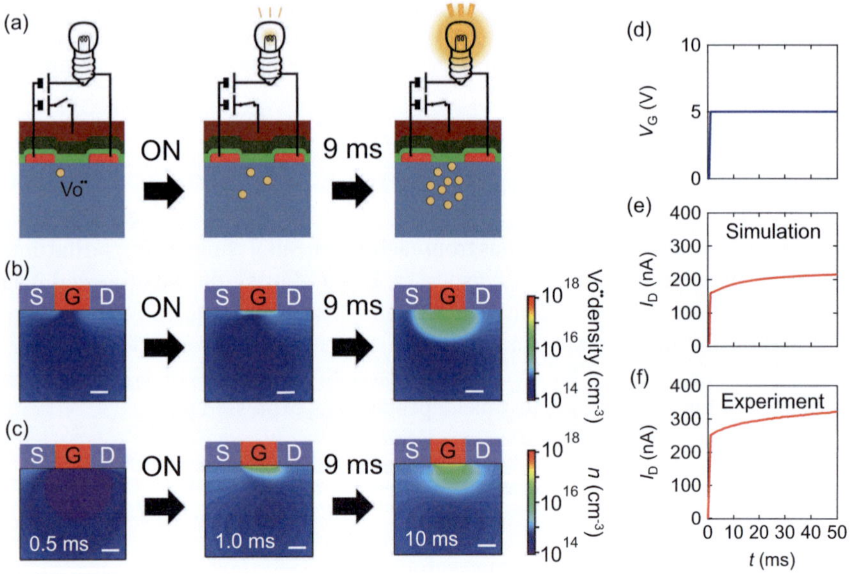

Fig. 8 Drift-diffusion mechanism of leaky integration. (**a**) Schematic illustration of the $V_O^{\cdot\cdot}$ distribution and its time evolution after the FET is switched on. (**b**) $V_O^{\cdot\cdot}$ density and (**c**) electron density n calculated from finite element analysis. The scale bars in (**b**) and (**c**) are 1 μm. (**d**)–(**f**) Comparison of the time evolution of I_D obtained from finite element analysis and experimental data when a step-wise voltage is applied as V_G. Adopted from [5]

to the applied electric field. When a positive gate voltage is applied, electrons accumulate in the transistor channel through conventional field-effect gating [57]. Simultaneously, positively charged oxygen vacancies drift deeper into the substrate due to the repulsive force exerted by the gate voltage. This results in a non-equilibrium distribution, with a reduced vacancy density near the $SrTiO_3$ surface. To compensate for this reduction, additional vacancies are generated and injected from the surface through oxygen exchange mechanisms.

Several potential sources of oxygen vacancies exist. One possibility is the oxygen-deficient regions near the source and drain electrodes, resulting from the fabrication process. In the $SrTiO_3$ transistor, these electrodes were formed by depositing titanium metal onto the $SrTiO_3$ surface. This process creates a significant amount of oxygen vacancies through oxygen gettering beneath the Ti electrodes, turning these regions into "reservoirs" that supply vacancies to the channel. Another potential source is electrochemical water splitting [27]. Although water is not intentionally introduced into the device structure, environmental moisture could contribute to water splitting, thus reducing $SrTiO_3$ when under a more negative potential. Furthermore, protons can migrate across the gate insulator stack and reduce $SrTiO_3$ when injected onto its surface [58–62].

In any case, the total number of oxygen vacancies increases due to vacancy drift-diffusion and vacancy generation within the channel. Since these vacancies act as electron donors, electron density also increases. The resulting I_D reflects this increase through the Drude model, $I_D \propto \sigma_e E = en\mu E$, where σ_e, E, n, and μ represent electrical conductivity, electric field, electron density, and electron mobility, respectively. Therefore, the slow drift-diffusion of oxygen vacancies leads to the observed leaky integration behavior.

When the gate voltage is removed, thermal diffusion restores the oxygen vacancies to their original distribution. This recovery process is accelerated by applying a negative voltage to the gate electrode.

The conceptual model illustrated in Fig. 8(a) is evaluated using numerical simulations as follows. The electrostatic potential in $SrTiO_3$ is described by Poisson's equation, which depends on the electron density n and oxygen vacancy density N_{vo}:

$$\nabla \cdot (-\varepsilon \nabla \phi) = e(-n + 2N_{vo}). \tag{7}$$

The gradient of ϕ drives the drift-diffusion of oxygen vacancies according to the diffusion equation:

$$\frac{\partial N_{vo}}{\partial t} = \nabla \cdot \left[D_{vo} \nabla N_{vo} + \frac{2eD_{vo}}{k_B T} N_{vo} \nabla \phi \right]. \tag{8}$$

The same potential ϕ, combined with the quasi-Fermi level E_F, determines the electron density n via the Fermi-Dirac distribution:

$$n = \int_{-\infty}^{\infty} D_n(\mathcal{E} + \phi) \frac{1}{1 + \exp\left(\frac{e(\mathcal{E} - E_{\mathrm{F}})}{k_B T}\right)} d\mathcal{E}, \tag{9}$$

where D_n is the density of the states in the $SrTiO_3$ conduction band. Only electron contributions are considered here, as hole contributions are negligible due to the large bandgap of $SrTiO_3$. Additionally, the E_{F} of $SrTiO_3$ is known to lie near the conduction band edge. The electrical current density J_n is calculated from the drift-diffusion of electrons, driven by the gradient of E_{F}:

$$J_n = en\mu \nabla E_{\mathrm{F}}, \tag{10}$$

along with electron conservation:

$$e\frac{\partial n}{\partial t} = \nabla \cdot J_n. \tag{11}$$

Since analytical solutions for Eqs. (7) through (11) are unavailable, they are solved numerically.

The dynamics of oxygen vacancies were investigated using the V_{G} ramping scheme shown in Fig. 8(d) through finite element analysis implemented in COMSOL Multiphysics (COMSOL, Inc.) [5]. The calculations were performed in a two-dimensional geometry for simplicity, assuming uniformity along the third dimension.

At $t = 0.5$ ms, oxygen vacancies are depleted from the channel region due to the work function difference between $SrTiO_3$ and Au (Fig. 8(b)). When V_{G} is ramped to 5 V at $t = 1.0$ ms, oxygen vacancies accumulate beneath the top surface in the channel region but remain confined. Over the next 9 ms, they gradually spread deeper into the $SrTiO_3$ substrate due to drift-diffusion driven by the electric field of the applied gate voltage. This spread extends over several microns, facilitated by the spatially extended electric field supported by the large dielectric constant of $SrTiO_3$ (Fig. 9 (d)).

The spatial distribution of electrons is shown in Fig. 8(c), which closely follows the oxygen vacancy distribution. This correlation can be better understood by examining the band structure along the vertical centerline of the device, as illustrated in Fig. 9(a)–(c).

Initially, the conduction band of $SrTiO_3$ exhibits upward bending, leading to electron depletion in the channel region (Fig. 9(a)). After V_{G} is applied at $t = 1.0$ ms, the conduction band bends downward, accumulating electrons near the $SrTiO_3$ surface (Fig. 9(b)). Notably, the quasi-Fermi level (E_{F}) also bends slightly downward toward the surface, indicating electron drift-diffusion toward the surface. This movement counteracts the drift-diffusion of oxygen vacancies away from the surface. At $t = 10$ ms, the conduction band bending becomes less pronounced, leading to a broader electron distribution compared to $t = 1.0$ ms (Fig. 9(c)).

The corresponding electrical current density is also distributed deeper into the $SrTiO_3$ substrate due to the extended electron distribution. Integrating this current

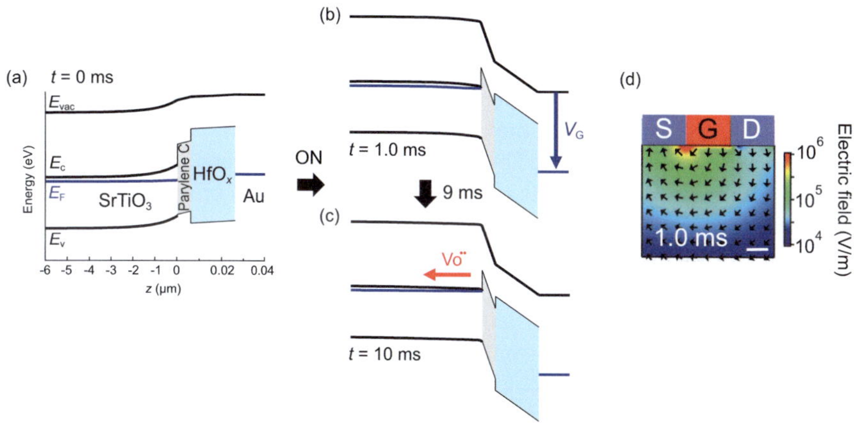

Fig. 9 Band structure of $SrTiO_3$ (**a**) before V_G application, (**b**) just after the FET is switched on, and (**c**) after drift-diffusion of $V_O^{\cdot\cdot}$. (**d**) Spatial distribution of electric fields just after the FET is switched on. The colors and arrows indicate the magnitude and direction of the electric fields, respectively. The scale bar is 1 μm. Adopted from [5]

density yields I_D, which shows leaky integration behavior (Fig. 8(e)), consistent with experimental observations (Fig. 8(f)).

5.4 *Advantages of Oxygen Vacancy Drift-Diffusion*

What is special about the oxygen vacancy drift-diffusion mechanism? There are several advantages in the perspective of application to slow electronics. The mechanism allows to achieve the long time-scale relatively easily with small power consumption and in small device size. The time-scale of the $SrTiO_3$ FET is longer than many other devices listed in Table 1 and fall in the range close to that of biological neurons. Since the time-scale is dominated by the motion of speed specific to the type of ions and the host material, the long time-scale would be maintained even if the device size is shrunk down for large scale integration. This invariability contrasts to many other mechanisms; e.g. time-scale scales with a capacitor size. The power consumption is as small as 500 pW because of the negligible current (< 30 pA) required to drive oxygen-vacancy drift-diffusion. Furthermore, the ground state of the device does not change over repeated voltage cycles because the drift-diffusion is a reversible process. We have experimentally tested over 10^8 cycles, which does not show signs of degraded device characteristics [5]. This robust endurance is an another advantage compared to the devices that utilize ferroelectrics or electrochemical forming.

5.5 Implementing Leaky-Integrate-and-Fire Neuron

The $SrTiO_3$ drift-diffusion FET efficiently realizes the leaky-integration function. To emulate the leaky-integrate-and-fire behavior of biological neurons, a "fire" function must be added. This was achieved by incorporating a simple external circuit with the $SrTiO_3$ FET, as shown in Fig. 10(a) [55]. The circuit consists of three main blocks: a current amplifier, a pulse generator, and a reset mechanism.

The current amplifier utilizes a bipolar junction transistor (BJT) T1, which senses the small drain current (I_D) from the $SrTiO_3$ FET at its base terminal and outputs an amplified current at the collector terminal. This amplified current is then converted to a voltage (V_{LI}) through a resistor (R1) connected to the collector.

The pulse generation block compares V_{LI} with the threshold voltage of inverter U2, triggering the reset circuitry when the threshold is exceeded. T2, along with R4, R5, and C1, forms a positive feedback that stabilizes the inputs of U1 and U2 to a logical high state, producing a voltage pulse at the output (V_{out}). Simultaneously, the reset circuitry T3 pulls the gate terminal of the $SrTiO_3$ FET to a negative voltage, resetting the oxygen vacancy distribution to its initial state.

The duration of the reset is governed by a timing circuit comprising C1, R1, R2, and R3. The gradual discharge of C1 eventually drives the inputs of U1 and U2 to a logical low state. The output spike duration is set shorter than the reset period by adjusting the threshold voltage of U1 relative to that of U2. Additionally, C1 and R3 function as a low-pass filter to eliminate glitches caused by the displacement current of the $SrTiO_3$ FET during switching.

Figure 10(b) illustrates the output waveforms when the circuit receives a pulse train. At an input frequency 50 Hz, V_{LI} leaky-integrates the input pulses until it

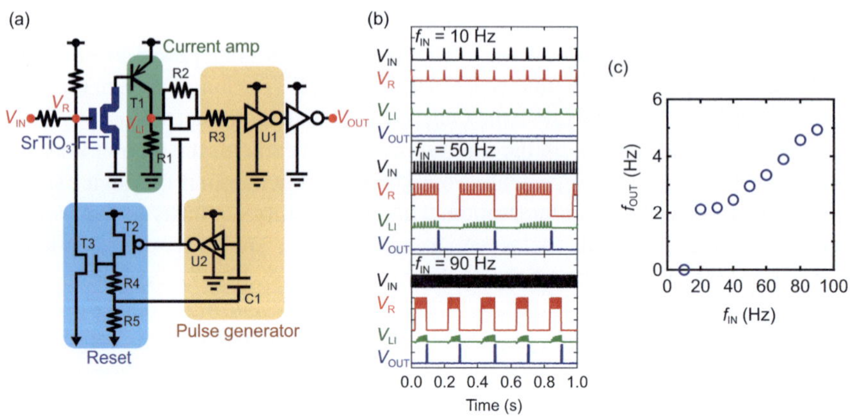

Fig. 10 Leaky-integrate-and-fire neurons constructed with the $SrTiO_3$ FET and an external circuit. (**a**) Schematic of the circuit. (**b**) Output waveforms of the circuit when the input V_{IN} is excited by pulses at different frequencies. (**c**) Output fire pulse frequencies as a function of input pulse frequencies. Adopted from [55]

reaches a threshold, triggering a spike at the V_{out} terminal, thus demonstrating the leaky-integrate-and-fire function. Concurrently, the gate voltage (V_G) of the $SrTiO_3$ FET is set to a negative value, redistributing the oxygen vacancies back to their initial state.

The interval between output spikes decreases as the input frequency increases; for example, 90 Hz, spikes occur more frequently than 50 Hz, as V_{LI} reaches the threshold more rapidly. Conversely, no spikes are generated at an input frequency 10 Hz, as V_{LI} never reaches the threshold.

Figure 10(c) summarizes the relationship between the output frequency (f_{out}) and input frequency (f_{in}). For $f_{in} > 20$ Hz, f_{out} increases monotonically with f_{in} until saturating at $f_{out} \sim 5$ Hz.

5.6 Constructing a Four-Neuron Reservoir

By combining multiple $SrTiO_3$-FET-based artificial neurons, an artificial neural network can be constructed. Such neural networks are physical systems that exhibit nonlinear and complex dynamics through interacting neurons. Therefore, they serve as a practical platform for implementing reservoir computing, as discussed in Chap. 2. The $SrTiO_3$-FET-based neural network is particularly attractive for reservoir computing applications due to its time scale, which is close to that of human-oriented signals, and its energy efficiency.

To construct this neural network, four leaky-integrate-and-fire (LIF) neurons were interconnected as shown in Fig. 11(a), (b). The leaky-integration function was

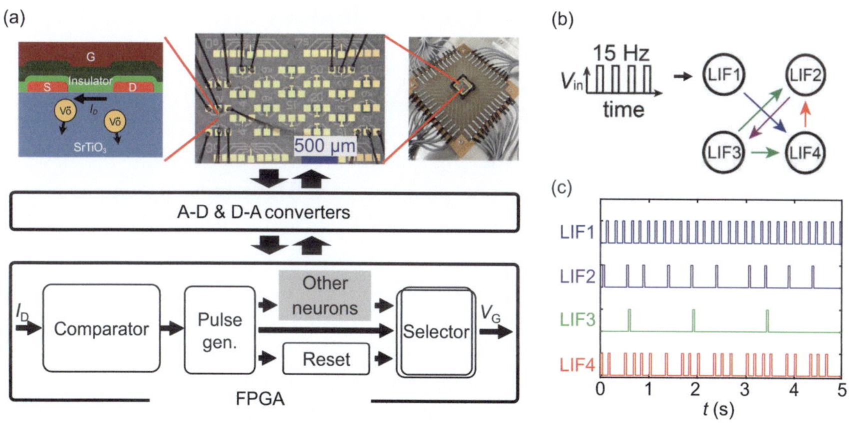

Fig. 11 Four-neuron reservoir constructed with $SrTiO_3$ FET-based leaky-integrate-and-fire neurons. (**a**) Device photograph and block diagrams of the firing functionality implemented by FPGA. (**b**) Schematic of the interconnected neuron circuit. (**c**) Output waveforms of neurons when LIF1 is excited 15 Hz pulses. Adopted from [55]

realized using real $SrTiO_3$ drift-diffusion FETs, while the firing functionality was implemented using an FPGA-based digital hardware platform[6] along with analog-to-digital (A/D) and digital-to-analog (D/A) converters for simplicity.

In the experiment, neuron LIF1 was driven by 15 Hz pulse train, and the outputs from all four neurons were monitored. Neurons LIF2 and LIF4 exhibited spikes with non-uniform intervals, as shown in Fig. 11(c). This indicates that the neural network transformed the periodic input pulses into irregular output spikes, demonstrating the basic functionality of a reservoir that nonlinearly transforms signals in the space-time domain [55].

5.7 Summary of Oxygen Vacancy Drift-Diffusion Transistor

Oxygen vacancy drift-diffusion field-effect transistor can realize leaky-integration behavior with single device. The long timescale is due to the drift–diffusion of oxygen vacancies acting as mobile donors under an applied electric field. Unlike conventional MOSFET devices, this FET shows conductance changes over long timescales (on the order of one second) with much lower energy consumption because the leaky integration occurs electrostatically with negligible net electrical currents or Joule heating. Timescale adjustability may be possible through electrostatic engineering, and the concept can be extended to other oxides. The long timescales are beneficial for building basic component of energy-efficient neuromorphic computing processing human interactive signals.

6 Proton Devices with Short-Term Memory

6.1 Introduction of Proton Devices

A highly effective method for achieving extended time scales in scalable devices involves leveraging resistance change phenomena [63], as they are characterized by their lengthy relaxation time constants. Among these mechanisms, ion transport in solid-state systems is particularly promising due to its excellent controllability through electrochemical processes [64]. Protons (H^+), in particular, exhibit remarkably high mobility at room temperature across various solids, thanks to their small ionic radius [65, 66]. This feature distinguishes them from other ions such as Li^+, Cu^{2+}, Ag^+, and O^{2-}. Additionally, while certain metal ions face challenges with contamination that render them incompatible with Complementary Metal-Oxide-Semiconductor (CMOS) technology, protons offer a significant advantage by being largely CMOS-compatible and capable of producing highly consistent changes in electrical conductivity.

---

[6] FPGA-related topics are discussed in Chap. 6.

Traditional protonic devices have predominantly employed a three-terminal design [67–77], where protons are introduced or removed from the channel by applying a gate voltage exceeding 1 V. This process modulates the channel's electronic conductivity through electron or hole doping. In contrast, two-terminal configurations, despite their simpler design, are less commonly utilized. Moreover, most existing two-terminal protonic devices have focused on lateral ion motion rather than vertical ion movement [78–82], which holds significant potential for enhanced control and scalability [83, 84]. Leveraging vertical proton dynamics offers the possibility of creating volatile electrical conductivity with tunable time scales, akin to the behavior of biological neurons and synapses. Such transient changes are also critical for Echo State Networks, a recently explored compact neural network architecture designed for real-time learning and inference [85, 86].

In this research, we developed scalable vertical two-terminal protonic devices capable of exhibiting volatile resistance changes over time scales ranging from seconds to minutes [26]. These resistance variations, driven by proton dynamics, were employed as a form of short-term information storage to demonstrate hardware for ESN algorithm. The findings highlight an efficient approach to real-time data processing, relying on the short-term retention of information through the slow relaxation behavior of protons, rather than utilizing high-retention memory as seen in traditional electronic systems.

6.2 *Fabrication of Protonic Device*

We constructed a two-terminal protonic device featuring a Pt/WO_3/Pd structure on a Si/SiO_2 substrate (Fig. 12a, b), characterized by well-defined interfaces with no detectable interdiffusion, as illustrated in Fig. 12c–f. The top layer, Pd, functions as a hydrogen storage material, while WO_3 serves as a proton conductor and a wide bandgap semiconductor that exhibits electronic conductivity upon proton doping [87]. The WO_3 layer, deposited via Pulsed Laser Deposition (PLD) at room temperature, was confirmed to be amorphous through analyses using X-Ray Diffraction (XRD) and Nano Beam Electron Diffraction (NBD).

The Pt/WO_3 interface functions as a hydrogen-blocking electrode because hydrogen cannot dissolve into the Pt layer at the lower electrode, allowing for straightforward device operation through hydrogen transfer between Pd and WO_3. This design enables volatile electrical conductivity that reverts to its initial state once the applied voltage is removed. Additionally, the electrode potential of slightly hydrogenated Pd on the upper electrode closely matches that of hydrogenated WO_3, facilitating low-voltage proton control at the WO_3/Pd interface [88, 89]. Moreover, the relaxation time constant of protons in the 120 nm-thick amorphous WO_3 layer, estimated to be around 10 s based on the diffusion constant ($D \sim 10^{-11}$-10^{-10} cm^2/s), allows the Pt/WO_3/Pd structure to support resistance changes with time constants of 10 s or longer, aligning with the goals of this study [90]. Notably, the top Pd electrode also

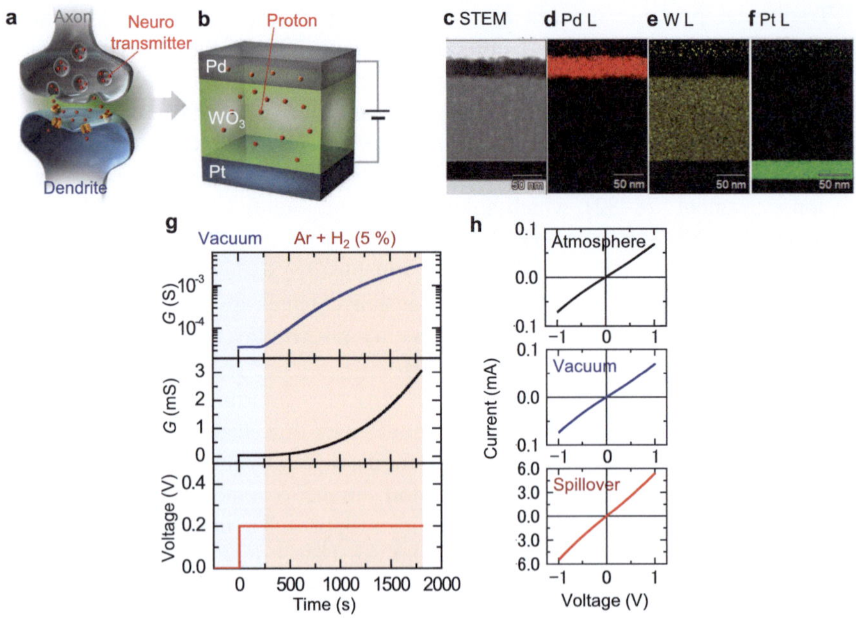

Fig. 12 **a** Schematic representation of biological synapse between the pre-neuron (axon) and the post-neuron (dendrite). **b** The design of two-terminal protonic device comprising $Pt/WO_3/Pd$ trilayer. **c** Cross-sectional HAADF-STEM image of $Pt/WO_3/Pd$ trilayer. **d–f** Corresponding EDX elemental mappings which show the spatial distribution of Pd, W, and Pt, respectively. **g** Conductance as a function of time with a measurement voltage of 0.2 V in Ar + H_2 (5%) atmosphere. **h** $I - V$ curves in the air (as-deposited), in the vacuum (before spillover), and in Ar + H_2 (5%) atmosphere after spillover [26]

acts as a catalyst for H_2 molecules, enabling the initial introduction of hydrogen into the device through H_2 exposure [91].

6.3 *Proton Introduction by Spillover*

For the spillover measurements (Fig. 12g), experiments were performed within a vacuum chamber, where a gas mixture of Ar and H_2 (5%) was kept at a pressure of 1 atm. The flow rate of the gas was regulated at 200 ml/min using a mass flow controller. The conductance modulation measurements under applied voltage (Fig. 13) were conducted in open air with a relative humidity of around 45%.

The device was initially placed in a vacuum chamber, and at 200 s, an Ar + H_2 (5%) gas mixture was introduced, as depicted in Fig. 12g. Following this, the electrical conductance increased significantly over several orders of magnitude, with a 0.2 V voltage applied to the top Pd electrode at 0 s. This change in conductance is attributed to hydrogen spillover, where hydrogen molecules break down on the

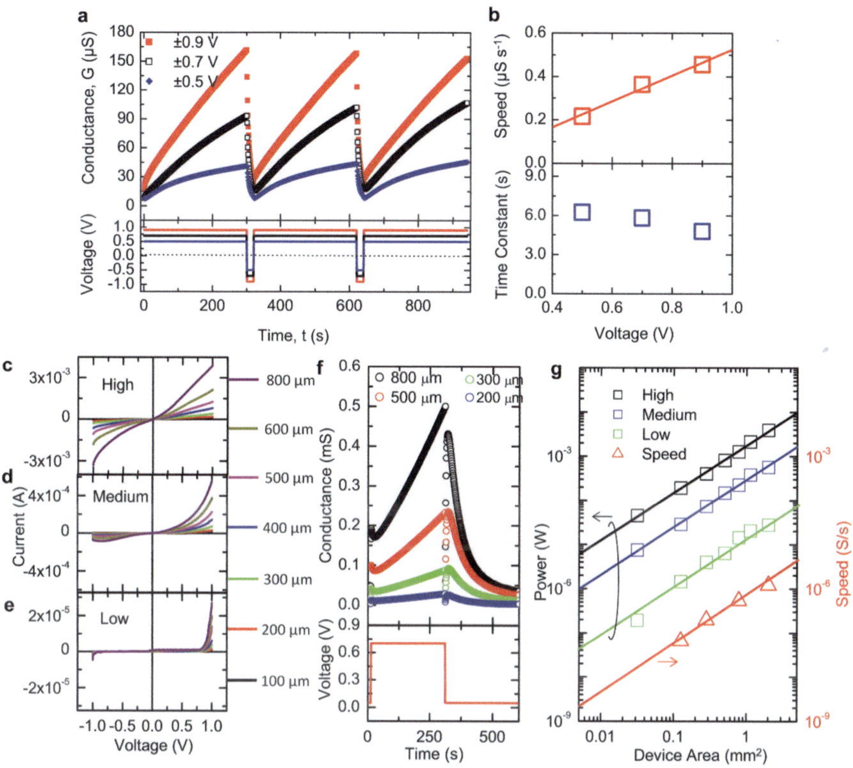

Fig. 13 **a** Conductance change by positive and negative voltage application (±0.5 V, ±0.7 V, and ±0.9 V). Here, positive voltages were applied for 300 s, and negative voltages were applied for 10 s. **b**, The speed of conductance increase under positive voltage is evaluated by linear fitting, and the time constant under the negative voltage was evaluated by exponential fitting. **c**–**e** $I - V$ curves for various device diameters (800 μm down to 100 μm) with **c** high, **d** medium, and **e** low proton contents, which were controlled by post-annealing (6, 12, and 18 min) at 120 °C. f, Conductance change by pulse voltage (0.7 V, 300 s) for various device diameters. **g** Log-log plot of area-dependent power dissipation at 1 V and corresponding linear fitting for three different proton contents in **c**–**e**. The area dependence of the speed in conductance change is also plotted, which was evaluated from the linear slope in **f** [26]

Pd surface and diffuse into the WO_3, making it more electrically conductive [91]. The *I*-*V* data presented in Fig. 12h demonstrated Ohmic behavior in ambient air, in vacuum, and in the Ar + H_2 atmosphere, indicating that the conductance is primarily determined by the bulk properties of WO_3, rather than its interface. The minimal effect at the interface suggests the formation of an electric double layer between the protons in WO_3 and the counterelectrons in Pd [92, 93], which prevents the depletion of WO_3 and the formation of a Pd/WO_3 Schottky junction due to the work function difference (5.2 eV for Pd [94] and 4.6 eV for WO_3) [95].

6.4 Conductance Modulation

Once spillover reaches saturation, the conductivity of the WO_3 layer becomes so high that most of the applied voltage is dropped across the Pt bottom electrode, rather than across the WO_3 itself. To ensure that voltage is applied directly to the WO_3 layer and to enable control over the protons, we carried out thermal annealing under moderate conditions (120°C for 12 min), which reduced the proton concentration in the WO_3 and increased its resistance. This allowed the voltage to be effectively applied to the WO_3 layer, and electrical conductance could then be modulated by applying voltages of ± 0.5 V, ± 0.7 V, and ± 0.9 V to the Pd top electrode, as shown in Fig. 13a. These results suggest that protons within the two-terminal device can be dynamically controlled via voltage application. Similar to the spillover phenomenon, a positive voltage applied to the Pd electrode led to an increase in conductance, while a negative voltage resulted in a decrease. Interestingly, the increase in conductance with positive voltage occurred gradually over a period of more than 100 s, whereas the decrease in conductance with negative voltage was much quicker, taking only around 10 s. The voltage dependence of the conductance change is summarized in Fig. 13b, where the rate of conductance change was extracted from the conductance onset in Fig. 13a. As the applied voltage increased, both the magnitude and speed of the conductance change grew. The decrease in conductance followed an exponential decay, with the extracted time constant shown in Fig. 13b. This time constant showed little dependence on the applied voltage and was roughly consistent with the value ($\sim$7 s) derived from the diffusion constant in amorphous WO_3 ($D \sim 10^{-11}$-10^{-10} cm^2/s) [90] and the film thickness ($L = 120$ nm), using the formula $\tau = \frac{L^2}{(2D)}$.

The observed asymmetric conductance variation with voltage polarity might be due to the higher hydrogen energy in WO_3 compared to the Pd surface, by approximately 0.22 eV per hydrogen atom [96, 97]. Since the behavior of individual hydrogen atoms is primarily governed by their energy (the enthalpy change during hydrogenation) and the energy barrier involved, the greater hydrogen energy in WO_3 may cause its hydrogenation process to occur more slowly than dehydrogenation.

6.5 Device Scaling

In many resistance-switching devices that rely on ionic motion, the resistance is controlled by localized conductive filaments, and this resistance does not scale with the device's area [98]. As a result, reducing the device size does not necessarily lower power consumption. To explore the scaling effect in protonic devices, we created devices with diameters ranging from 100 μm to 800 μm. Figure 13c-e show the current-voltage (I-V) curves, which vary with the device scale and three different thermal annealing conditions (6, 12, and 18 min at 120 °C), resulting in high, medium, and low proton concentrations, respectively. In all cases, the current was proportional to the device's area, suggesting a fairly uniform proton distribution across the device after thermal treatment. The power dissipation at 1 V, plotted against the device area in Fig. 13g, clearly demonstrates that reducing the device size can significantly decrease power consumption. Additionally, Fig. 13f illustrates the area dependence of conductance change when the voltage was varied between 0.05 V and 0.7 V. The conductance change was reversible, and its rate was proportional to the device area (shown on the "Speed" axis in Fig. 13g), indicating that the conductance change, driven by voltage application, is governed not by localized conductive filaments but by the uniform motion of protons within the device.

6.6 *Leaky Integration*

As shown in Fig. 13a, the resistance change rate exhibits asymmetry, with resistance decreasing gradually at positive voltages and increasing rapidly at negative or zero voltages. In other words, a brief negative voltage pulse causes a rapid increase in resistance, while a prolonged positive voltage leads to a slow relaxation. This behavior can be exploited to implement a well-known function in spike signal processing–namely, leaky integration.

Leaky integration of input spikes results in short-term averaging, essentially converting spikes into analog signals, which is a fundamental aspect of real-time biological system operations. To demonstrate this concept, the device was maintained at 0.5 V, and a downward pulse of 0.05 V was applied for 5 s every 50 s, as illustrated in Fig. 14. Each downward pulse caused an increase in resistance, but as the resistance increased, the leak rate also grew. Consequently, the resistance reached a saturation point, as shown in Fig. 14, which is influenced by the frequency of the input spike signal [99]. The leak rate is governed by the relaxation time constant of protons, which defines the timescale for short-term information storage and, therefore, the time scale over which the device can process information.

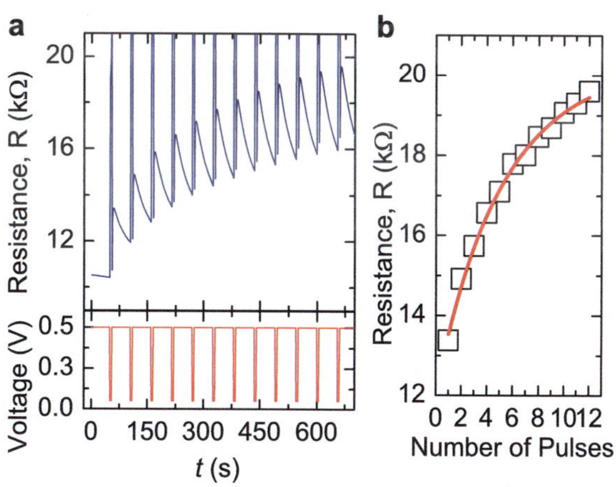

Fig. 14 **a** Leaky integration with $\Delta t = 50\,\mathrm{s}$. **b** The resistance values just after the pulse application in **a** are plotted as a function of pulse number [26]

6.7 Application to the Circuit for Echo State Network

Leaky integration, with a relaxation time spanning several seconds, serves as a form of short-term memory, which plays a crucial role in biological information processing. This approach will be demonstrated as effective for real-time prediction of time-series data, such as blood glucose levels, that fluctuates on a minute-by-minute basis. Blood glucose is a critical factor in preventing heart disease, typically rising after meals and falling due to insulin secretion [85]. However, when the body's regulation of blood glucose is impaired, abnormal levels can occur, leading to harm. As illustrated in Chap. 5, if blood glucose levels can be predicted ahead of time, and medication is administered when the predicted level exceeds a certain threshold, it could significantly help in preventing heart disease.

The prediction of blood glucose levels using protonic devices was modeled within an Echo State Network (ESN) framework. ESN is a type of recurrent neural network designed for time-series data processing, where the internal connection weights are fixed and only the output weights are learned [100, 101]. This feature makes ESN training straightforward, requiring minimal computational resources, often accomplished through linear regression or similar methods. As a result, ESN could enable real-time blood glucose prediction even on devices with limited processing power, such as wearables. In the analog implementation, each ESN node needs a short-term memory function in line with the algorithm's requirements (not long-term), and volatile devices with a time constant of several minutes are essential. This extended time constant is necessary because the short-term memory in the ESN should match the time scale of the input data. Therefore, as discussed further in the next section, protonic devices are used for this short-term memory function. The protonic device serves as the "R_{WO3}" device (in red) shown in Fig. 15c. Since the protonic device can leaky integrate the input spike signal into electrical conductance, it acts as short-term memory within the circuit of Fig. 15c. Each node in the ESN of Fig. 15b temporarily stores analog data in the protonic device and carries out nonlinear processing using analog CMOS circuits. Consequently, the entire network is capable of performing various inference tasks according to the ESN algorithm.

When attempting to implement the ESN algorithm in dedicated hardware, handling information as analog voltage or current values proves to be challenging. In contrast, biological systems suggest that spike signal frequencies offer a more straightforward method for encoding information. In fact, spike signals can be processed by basic analog CMOS circuits. In this study, we designed an ESN node consisting of a protonic device and an analog CMOS circuit, as illustrated in Fig. 15c. This circuit leaky integrates the input spike signal into the resistance of the protonic device (R_{WO3}), and outputs a spike signal whose frequency is directly proportional to the resistance value. A timing diagram of the circuit operation is shown in Fig. 15d (for more details, see Methods). The "saturation" section in Fig. 15c ensures the output saturates at high input spike frequencies (f_{IN}), while the "nonlinear" section causes the output spike frequency (f_{OUT}) to exhibit a nonlinear response to f_{IN}. As a result, the circuit in Fig. 15c can produce a hyperbolic tangent (tanh) input-output relation-

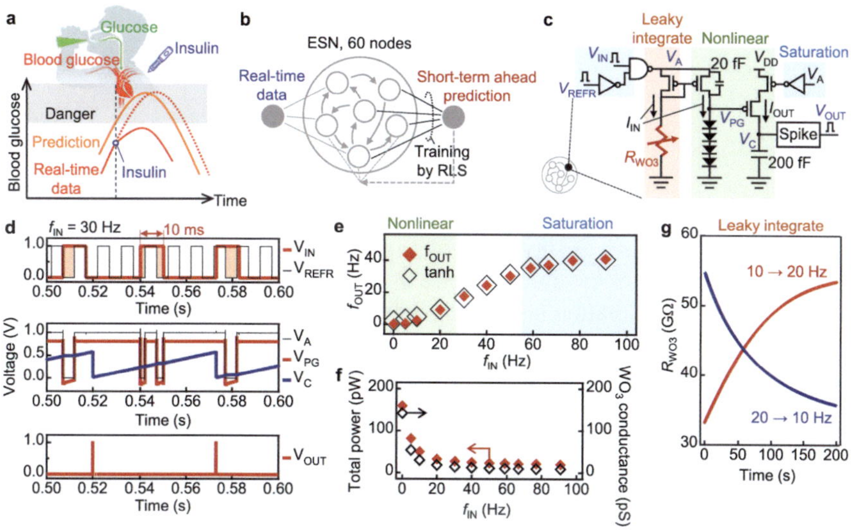

Fig. 15 **a** A schematic illustration of blood glucose level prediction. **b** The ESN that is used in the prediction task. **c** The protonic device-embedded analog CMOS circuit that was used as an ESN node. **d** A time chart of the circuit operation in **c**. The filtering by V_{REFR} achieves saturation characteristics, and V_{PG} nonlinearly couples I_{IN}^{-1} ($\propto R_{WO3}$) and I_{OUT} ($\propto f_{OUT}$). **e** Hyperbolic tangent input-output relationship of the ESN node. **f** Power consumption of the ESN node and the corresponding conductance of the protonic device under the steady state. **g** Relaxation characteristics of the protonic device resistance (R_{WO3}) for a sudden change of the input frequency (f_{IN}) [26]

ship, which is a key feature in ESNs, as shown in Fig. 15e. The power consumption of this ESN node can be minimized to below 160 pW, as depicted in Fig. 15f, assuming a scaled-down protonic device (see Methods). Therefore, even with 60 nodes, the total power consumption of the ESN can be as low as 10 nW, according to simulations using TSMC 180 nm technology.

The circuit simulation shown in Fig. 15c was conducted using Cadence's Spectre, utilizing the TSMC 180nm BCD process. A protonic device model, developed from experimental data, was integrated into the simulation through Verilog A code.

$$\frac{dG(t)}{dt} = \frac{G_s(V(t)) - G(t)}{\tau} \tag{12}$$

G(t) represents the conductance of the protonic device, $V(t)$ is the voltage applied to the Pd top electrode, $G_s(V)$ denotes the saturated conductance at a given voltage V, and τ is the relaxation time constant. The values of $G_s(V)$ and τ were estimated based on the experimental data shown in Fig. 13, as described below:

$$G_s(V) = G_0(1 + \exp(5V)) \tag{13}$$

$$\tau = 100\,s \text{ for } G_s\left(V\left(t\right)\right) > G\left(t\right) \tag{14}$$

$$\tau = 6\,s \text{ for } G_s\left(V\left(t\right)\right) < G\left(t\right) \tag{15}$$

G_0 represents the steady-state conductance value when no voltage is applied. For the simulation, we assumed $G_0 = 5$ pS, based on the scaling of the protonic device to approximately 1 μm^2, as shown in Fig. 13g.

Figure 15d illustrates the time sequence of the circuit in Fig. 15c. A spike voltage (V_{IN}) with a 10 ms duration is applied, and the common portion with a reference square wave (V_{REF}) is fed into the protonic device (V_A). V_{REF} sets the upper boundary for the input to the protonic device and plays a role in the saturation behavior. Each time V_A is applied to the protonic device, it undergoes leaky integration, gradually increasing the resistance value (R_{WO3}). When V_A is applied to R_{WO3} through a PMOS transistor, a current (I_{IN}), which is inversely proportional to R_{WO3}, flows. Subsequently, a current (I_{OUT}) that has an inverse relationship with I_{IN} flows into the spike generation circuit. Consequently, I_{OUT} is directly proportional to R_{WO3}, reflecting the leaky integration of the input signal. The conversion from I_{IN} to I_{OUT} is highly nonlinear, which contributes to the nonlinear characteristics. I_{OUT} then charges a small on-chip capacitor (200 fF), and whenever the capacitor voltage (V_C) reaches 0.5 V, an ultra-low-power spike generation circuit produces an output spike (V_{OUT}) and resets V_C to zero. Therefore, f_{OUT} is directly related to I_{OUT} and, in turn, to R_{WO3}. It is important to note that I_{OUT} is paused during V_A's application to the protonic device, which results in the saturation behavior of f_{OUT}. As a result, the relationship between f_{IN} and f_{OUT} in the steady state is nonlinear due to the I_{IN}-I_{OUT} conversion, and it exhibits saturation effects due to the filtering from the V_A signal and V_{REF}, leading to a hyperbolic tangent function as shown in Fig. 15e.

Crucially, the integrated input spike information is stored in R_{WO3} for several tens of seconds, making it usable for the ESN's prediction task. Additionally, the recurrent nature of the network causes this information to circulate internally, extending its retention even further [102]. As a result, the ESN can generate outputs based on past input information, ranging from recent to earlier data, without requiring explicit memory. In fact, the gradual change in R_{WO3} when f_{IN} experiences sudden shifts (Fig. 15g) demonstrates that past input information continues to affect R_{WO3} for an extended period. It's important to note that if the relaxation time of the protonic device were significantly shorter, the input information would fade too quickly, rendering the prediction of slow-changing time-series data, such as biomedical data, unfeasible.

6.8 Correspondence to Echo State Network Algorithm

The inputs to the ESN ($x(t)$ mg/dL) in the blood glucose prediction (Chap. 5) represent the 5-minute interval blood glucose measurements obtained from Continuous Glucose Monitoring (CGM) [103]. In this context, t refers to time in minutes. The data source is a study that evaluated continuous glucose sensor profiles in healthy, non-diabetic participants under 7 years of age. However, the analysis, content, and conclusions presented are solely those of the authors and have not been reviewed or endorsed by the Jaeb Center for Health Research. To reduce noise in the CGM data, the input signals were subjected to a low-pass filter with a cut-off frequency of (50 min)-1. The ESN state $\mathbf{r}\,(t)$ was updated according to the following equation.

$$\mathbf{r}\,(t+5) = (1-\alpha)\,\mathbf{r}\,(t) + \alpha tanh\left(\hat{w}\mathbf{r}\,(t) + s_{in}\mathbf{w_{in}}x\,(t+5)\right) \quad (16)$$

The input $x\,(t)$ was scaled by a factor of $s_{in} = 0.01$ and applied with input connection weights ($\mathbf{w_{in}}$), which were initialized with random values ranging from -1 to 1. The ESN architecture includes a recurrent network consisting of 60 nodes ($N = 60$), and the inter-node connection weights ($\hat{w}$, an N-by-N matrix) were randomly assigned values between -1 and 1, with a 10% probability of connection. This means that 90% of the entries in the $\hat{w}$ matrix were zero. At each time step (every 5 min), each node sums the inputs from all connected nodes and transforms this sum nonlinearly using a hyperbolic tangent function to produce $\mathbf{r}\,(t)$. The components of $\mathbf{r}\,(t)$ are then linearly combined using the output connection weights (w_out) to generate the ESN output, $y\,(t) = \mathbf{r}\,(t)\cdot\mathbf{w_{out}}$. The leak rate, α, was set to 0.9.

During training, the output weights $\mathbf{w_{out}}$ were updated using Recursive Least Squares (RLS) with a regularization constant of 0.0001, using future blood glucose data as the target signal. In the inference phase, the output weights $\mathbf{w_{out}}$ were held constant and not updated.

A model of the ESN node was developed using the input-output relationship from the analog CMOS circuit shown in Fig. 15e. By employing the function $f_{out} = F(f_{in})$ obtained in Fig. 15e, the nonlinear function used in the ESN update process was constructed as follows:

$$20\,(q+1) = F\,(20\,(p+1.75)) \quad (17)$$

Next, p and q are related by the equation $q = \tanh(p)$. Utilizing this $F\,(\cdot)$ function, we can initially express the ESN update equation without the leak term as follows:

$$\mathbf{r}\,(t+1) = tanh\left(\hat{w}\mathbf{r}\,(t) + s_{in}\mathbf{w_{in}}x\,(t+1) + s_{fb}\mathbf{w_{fb}}y\,(t)\right) \quad (18)$$

By setting $p = \left[\hat{w}\mathbf{r}\,(t) + s_{in}\mathbf{w_{in}}x\,(t+1) + s_{fb}\mathbf{w_{fb}}y\,(t)\right]_i$ and $q = [\mathbf{r}\,(t+1)]_i$ ($[\mathbf{A}]_i$ denotes the ith component of the vector $\mathbf{A}$), the ESN update equation can be written as follows:

$$[20\,(\mathbf{r}\,(t+1)+\mathbf{1})]_i = F\left(\left[\hat{w}\,(20\,(\mathbf{r}\,(t)+\mathbf{1})) + \mathbf{w_{in}}\,(20\,(s_{in}x\,(t+1)+1)) + \mathbf{b_{cont}}\right]_i\right) \quad (19)$$

In this case, $\mathbf{b_{cont}}$ represents a time-invariant input for each ESN node and is defined such that $\hat{w}\mathbf{1} + \mathbf{w_{in}} + \mathbf{b_{cont}} = 1.751$, where $\mathbf{1}$ is a vector of dimension N

with all components equal to 1. This can be reformulated by introducing a function $F(\mathbf{A})$ that applies the $F(\cdot)$ function to each element of vector $\mathbf{A}$, as shown below.

$$20\left(\mathbf{r}(t+1)+\mathbf{1}\right)=\mathbf{F}\left(\hat{w}\left(20\left(\mathbf{r}(t)+\mathbf{1}\right)\right)+\mathbf{w_{in}}\left(20\left(s_{in}x(t+1)+1\right)\right)+\mathbf{b_{cont}}\right) \tag{20}$$

Therefore, the ESN node can be realized using the CMOS analog circuit shown in Fig. 15c by incrementing the ESN signal by one, scaling it by a factor of 20, and representing this value through the spike frequency in the circuit. Finally, taking into account the relaxation time of the protonic device in Fig. 15g, the leak rate per time step (5 min) for the ESN is set to 0.9.

6.9 Summary of Proton Devices

In conclusion, we have successfully developed a vertical two-terminal protonic device capable of leaky integrating input signals over time scales ranging from seconds to minutes. Our device consists of an amorphous WO_3 layer, where conductance undergoes significant changes driven by voltage-controlled proton movement. The conductance alterations are volatile, with a relaxation time constant on the order of several tens of seconds, and this dynamic behavior can be utilized as short-term memory for real-time information processing. This concept can be demonstrated, for example, in a blood glucose prediction task, where past glucose levels are retained within the relaxation process of the protonic devices and their associated recurrent network, as shown in Chap. 5. This marks the first demonstration of real-time processing of information on a minute scale, leveraging the slow relaxation characteristics of materials as short-term memory. This achievement represents a significant advancement for neuromorphic computing, which contrasts with traditional high-speed computing by processing information in real time at the same pace as the surrounding environment.

References

1. M. Huang, M. Schwacke, M. Onen, del Alamo. J., J. Li, B. Yildiz, Adv. Mater. **35**, 2205169 (2023)
2. M.A. Zidan, J.P. Strachan, W.D. Lu, Nat. Electron. **1**, 22 (2018)
3. Y. Li, Z. Wang, R. Midya, Q. Xia, J.J. Yang, J. Phys. D: Appl. Phys. **51**, 503002 (2018)
4. G. Zhou, Z. Wang, B. Sun, F. Zhou, L. Sun, H. Zhao, X. Hu, X. Peng, J. Yan, H. Wang, W. Wang, J. Li, B. Yan, D. Kuang, Y. Wang, L. Wang, S. Duan, Adv. Electron. Mater. **8**, 2101127 (2022)
5. H. Inoue, H. Tamura, A. Kitoh, X. Chen, Z. Byambadorj, T. Yajima, Y. Hotta, T. Iizuka, G. Tanaka, I.H. Inoue, Adv. Mater. **37**, 2407326 (2025)
6. D. Drubach, *The Brain Explained* (Prentice Hall Health, Upper Saddle River, NJ, 2000)
7. G. Buzsáki, *Rhythms of the Brain* (Oxford University Press, New York, NY, 2006)
8. A.L. Hodgkin, A.F. Huxley, J. Physiol. **117**, 500 (1952)

9. D. Gray, W. S.M., *Foundations of cellular neurophysiology (3rd ed.)* (MIT Press, Cambridge, MA)
10. B. Wang, W. Ke, J. Guang, G. Chen, L. Yin, S. Deng, Q. He, Y. Liu, T. He, R. Zheng, Y. Jiang, X. Zhang, T. Li, G. Luan, H.D. Lu, M. Zhang, X. Zhang, Y. Shu, Front. Cell. Neurosci. **10**, 239 (2016)
11. D.H. O'Connor, S.P. Peron, D. Huber, K. Svoboda, Neuron **67**, 1048 (2010)
12. M. Shafi, Y. Zhou, J. Quintana, C. Chow, J. Fuster, M. Bodner, Neuroscience **146**, 1082 (2007)
13. D. Markovic, A. Mizrahi, D. Querlioz, J. Grollier, Nat. Rev. Phys. **2**, 499 (2020)
14. G. Indiveri, E. Chicca, R. Douglas, IEEE Transactions on Neural Networks **17**, 211 (2006)
15. G. Indiveri, B. Linares-Barranco, T. Hamilton, A. van Schaik, R. Etienne-Cummings, T. Delbruck, S.C. Liu, P. Dudek, P. Häfliger, S. Renaud, J. Schemmel, G. Cauwenberghs, J. Arthur, K. Hynna, F. Folowosele, S. Saïghi, T. Serrano-Gotarredona, J. Wijekoon, Y. Wang, K. Boahen, Front. Neurosci. **5**, 73 (2011)
16. A. Joubert, B. Belhadj, O. Temam, R. Héliot, The 2012 International Joint Conference on Neural Networks (2012)
17. X. Wu, V. Saxena, K. Zhu, S. Balagopal, IEEE Transactions on Circuits and Systems II: Express Briefs **62**, 1088 (2015)
18. S. Moradi, S.A. Bhave, R. Manohar, *in 2017 IEEE Symposium Series on Computational Intelligence* (IEEE, Honolulu, 2017)
19. S.A. Aamir, P. Muller, G. Kiene, L. Kriener, Y. Stradmann, A. Grubl, J. Schemmel, K. Meier, IEEE Transactions on Biomedical Circuits and Systems **12**, 1027 (2018)
20. X. Chen, Z. Byambadorj, T. Yajima, H. Inoue, I.H. Inoue, T. Iizuka, Appl. Phys. Lett. **122**, 074102 (2023)
21. X. Chen, T. Yajima, I.H. Inoue, T. Iizuka, Jpn. J. Appl. Phys. **61**, SC1051 (2022)
22. A. Sengupta, P. Panda, P. Wijesinghe, Y. Kim, K. Roy, Sci. Rep. **6**, 30039 (2016)
23. M.H. Wu, M.C. Hong, C.C. Chang, P. Sahu, J.H. Wei, H.Y. Lee, S.S. Shcu, T.H. Hou, *in 2019 Symposium on VLSI Technology* (IEEE, Kyoto, 2019)
24. S.O. Park, H. Jeong, J. Park, J. Bae, S. Choi, Nat. Commun. **13**, 2888 (2022)
25. U.Y. Won, Q. An V., S.B. Park, M.H. Park, V. Dam Do, H.J. Park, H. Yang, Y.H. Lee, W.J. Yu, Nat. Commun. **14**, 3070 (2023)
26. S.P. Pati, Y. Geng, S. Hamasuna, K. Fujiwara, T. Iizuka, H. Inoue, T. Yajima, Commun. Mater. **5**, 177 (2024)
27. H. Ohta, Y. Sato, T. Kato, S. Kim, K. Nomura, Y. Ikuhara, H. Hosono, Nat. Commun. **1**, 118 (2010)
28. J.Q. Yang, R. Wang, Z.P. Wang, Q.Y. Ma, J.Y. Mao, Y. Ren, X. Yang, Y. Zhou, S.T. Han, Nano Energy **74**, 104828 (2020)
29. J. Luo, L. Yu, T. Liu, M. Yang, Z. Fu, Z. Liang, L. Chen, C. Chen, S. Liu, S. Wu, Q. Huang, R. Huang, *in 2019 IEEE International Electron Devices Meeting* (IEEE, San Francisco, 2019)
30. Z. Wang, B. Crafton, J. Gomez, R. Xu, A. Luo, Z. Krivokapic, L. Martin, S. Datta, A. Raychowdhury, A.I. Khan, *in 2018 IEEE International Electron Devices Meeting* (IEEE, San Francisco, 2018)
31. Z. Wang, S. Joshi, S. Savel'ev, W. Song, R. Midya, Y. Li, M. Rao, P. Yan, S. Asapu, Y. Zhuo, H. Jiang, P. Lin, C. Li, J. Yoon, N.K. Upadhyay, J. Zhang, M. Hu, J.P. Strachan, M. Barnell, Q. Wu, H. Wu, R.S. Williams, Q. Xia, J.J. Yang, Nat. Electron. **1**(2), 137 (2018)
32. A.I. Khan, A. Keshavarzi, S. Datta, Nat. Electron **3**, 588 (2020)
33. S. Dutta, C. Schafer, J. Gomez, K. Ni, S. Joshi, S. Datta, Front. Neurosci. **14**, 634 (2020)
34. R. Cao, X. Zhang, S. Liu, J. Lu, Y. Wang, H. Jiang, Y. Yang, Y. Sun, W. Wei, J. Wang, H. Xu, Q. Li, Q. Liu, Nat. Commun. **13**(1), 7018 (2022)
35. G. Lee, H.J. Kim, E.J. Shin, S. Kim, T.I. Lee, B.J. Cho, IEEE Electron Device Letters **43**, 1375 (2022)
36. R. Waser, M. Aono, Nat. Mater. **6**, 833 (2007)
37. M.J. Lee, C.B. Lee, D. Lee, S.R. Lee, M. Chang, J.H. Hur, Y.B. Kim, C.J. Kim, D.H. Seo, S. Seo, U.I. Chung, I.K. Yoo, K. Kim, Nat. Mater. **10**, 625 (2011)

38. Z. Wang, S. Joshi, S.E. Savel'ev, H. Jiang, R. Midya, P. Lin, M. Hu, N. Ge, J.P. Strachan, Z. Li, Q. Wu, M. Barnell, G.L. Li, H.L. Xin, R.S. Williams, Q. Xia, J.J. Yang, Nat. Mater. **16**(1), 101 (2017)
39. X. Zhang, W. Wang, Q. Liu, X. Zhao, J. Wei, R. Cao, Z. Yao, X. Zhu, F. Zhang, H. Lv, S. Long, M. Liu, IEEE Electron Device Letters **39**, 308 (2018)
40. S. Choi, S.H. Tan, Z. Li, Y. Kim, C. Choi, P.Y. Chen, H. Yeon, S. Yu, J. Kim, Nat. Mater. **17**, 335 (2018)
41. D. Lee, M. Kwak, K. Moon, W. Choi, J. Park, J. Yoo, J. Song, S. Lim, C. Sung, W. Banerjee, H. Hwang, Adv. Electron. Mater. **5**(9), 1800866 (2019)
42. S. Kumar, R.S. Williams, Z. Wang, Nature **585**, 518 (2020)
43. Q. Duan, Z. Jing, X. Zou, Y. Wang, K. Yang, T. Zhang, S. Wu, R. Huang, Y. Yang, Nat. Commun. **11**(1), 3399 (2020)
44. C. Du, F. Cai, M.A. Zidan, W. Ma, S.H. Lee, W.D. Lu, Nat. Commun. **8**, 2204 (2017)
45. C.D. Wright, P. Hosseini, J.A.V. Diosdado, Adv. Funct. Mater. **23**(18), 2248 (2013)
46. T. Tuma, A. Pantazi, M. Le Gallo, A. Sebastian, E. Eleftheriou, Nat. Nanotechnol. **11**, 693 (2016)
47. P. Stoliar, J. Tranchant, B. Corraze, E. Janod, M.P. Besland, F. Tesler, M. Rozenberg, L. Cario, Adv. Funct. Mater. **27**, 1604740 (2017)
48. W. Yi, K.K. Tsang, S.K. Lam, X. Bai, J.A. Crowell, E.A. Flores, Nat. Commun. **9**, 4661 (2018)
49. T. Yajima, T. Nishimura, A. Toriumi, in *2018 IEEE Symposium on VLSI Technology* (2018), pp. 27–28
50. O.N. Tufte, P.W. Chapman, Phys. Rev. **155**, 796 (1967)
51. H.P.R. Frederikse, W.R. Hosler, Phys. Rev. **161**, 822 (1967)
52. A. Kalabukhov, R. Gunnarsson, J. Börjesson, E. Olsson, T. Claeson, D. Winkler, Phys. Rev. B **75**, 121404 (2007)
53. P. Stoliar, A. Schulman, A. Kitoh, A. Sawa, I.H. Inoue, *in 2017 IEEE International Electron Devices Meeting* (IEEE, San Francisco, 2017)
54. N. Kumar, A. Kitoh, I.H. Inoue, Sci. Rep. **6**, 25789 (2016)
55. H. Inoue, H. Tamura, A. Kitoh, X. Chen, Z. Byambadorj, T. Yajima, Y. Hotta, T. Iizuka, G. Tanaka, I.H. Inoue, *in 2023 IEEE Symposium on VLSI Technology and Circuits* (IEEE, Kyoto, 2023)
56. S.M. Sze, M.K. Lee, *Semiconductor Devices: Physics and Technology* (John Wiley & Sons, New York, NY, 2012)
57. H. Inoue, H. Yoon, T.A. Merz, A.G. Swartz, S.S. Hong, Y. Hikita, H.Y. Hwang, Appl. Phys. Lett. **114**, 231605 (2019)
58. K.D. Kreuer, Chem. Mater. **8**, 610 (1996)
59. L.Q. Zhu, C.J. Wan, L.Q. Guo, Y. Shi, Q. Wan, Nat. Commun. **5**, 3158 (2014)
60. L.Q. Guo, H. Han, L.Q. Zhu, Y.B. Guo, F. Yu, Z.Y. Ren, H. Xiao, Z.Y. Ge, J.N. Ding, A.C.S. Appl, Mater. Interfaces **11**, 28352 (2019)
61. R.D. Nikam, J. Lee, W. Choi, W. Banerjee, M. Kwak, M. Yadav, H. Hwang, Small **17**, 2103543 (2021)
62. M. Onen, N. Emond, J. Li, B. Yildiz, J.A. del Alamo, Nano Lett. **21**, 6111 (2021)
63. J. Yang, D. Strukov, D. Stewart, Nature nanotechnology **8**, 13 (2013)
64. T. Marukame, K. Mizushima, K. Nomura, Y. Nishi, IEEE, J. Elec. Dev. Soc **11**, 602 (2023)
65. T. Yajima, AIP Adv **8**, 115133 (2018)
66. Z. Zhang, Nature **553**, 68 (2018)
67. X. Yao, Nat. Commun **11**, 3134 (2020)
68. H.M. Huang, Adv. Mater. **31**, 1803849 (2019)
69. Y. Liu, Adv. Mat **27**, 5599 (2015)
70. Y. Yang, A.C.S. Appl, Mater. Interfaces **8**, 30281 (2016)
71. M. Onen, Nano Lett. **21**, 6111 (2021)
72. J.T. Yang, Adv. Mater **30**, 1801548 (2018)
73. Y. Burgt, Nat. Mater **16**, 414 (2017)

74. P. Gkoupidenis, Nat. Comm **8**, 15448 (2017)
75. X. Ji, Nat. Comm **12**, 2480 (2021)
76. C. Wu, Nat Comm **8**, 752 (2017)
77. Z. Hemmatian, J. Mater. Chem. C **3**, 6407 (2015)
78. E. Josberger, Adv. Mater **26**, 4986 (2014)
79. L. Glasser, Chem Rev **75**, 21 (1975)
80. H. Morgan, R. Pethig, G. Stevens, J. Phys. E -Sci. Instrum **19**, 80 (1986)
81. Y. Deng, Sci. Rep **3**, 2481 (2013)
82. U. Sidik, A. Hattori, R. Rakshit, S. Ramanathan, H. Tanaka, A.C.S. Appl, Mater. Int **12**, 54955 (2020)
83. Z. Wang, H.M. Huang, X. Guo, Adv. Electron. Mater **7**, 2001243 (2021)
84. S. Yamaguchi, Electrochemistry **82**, 817 (2014)
85. E. Ackerman, Bull. Math. Biophysics **27**, 21 (1965)
86. T. Yajima, Sci. Rep **12**, 1150 (2022)
87. X. Song, Fuel Cell Operation. Crystals **11**, 1595 (2021)
88. A. Pundt, R. Kirchheim, Annu. Rev. Mater. Res **36**, 555 (2006)
89. J.V. Kerkhof, W. Olthuis, P. Bergveld, M. Bos, Sens. Act. B **3**, 129 (1991)
90. A. Jonsson, Journal of The Electrochemical Society **152**, A377 (2005)
91. J. Benson, H. Kohn, M. Boudart, J. Catal **5**, 307 (1966)
92. S. Bredikhin, T. Hattori, M. Ishigame, Phys. Rev. B **50**, 2444 (1994)
93. T. Yajima, Nature Commun **6**, 6759 (2015)
94. S. Halas, T. Durakiewicz, P. Mackieicz, Surf. Sci **555**, 43 (2004)
95. M. Grätzel, Nature **414**, 338 (2001)
96. Y. Xi, Q. Zhang, H. Cheng, J. Phys. Chem **118**, 494 (2014)
97. M. Schwarzer, J. Phys. Chem. C **126**, 14500 (2022)
98. J. Yang, Nature Nanotech **3**, 429 (2004)
99. P. Pillai, M.D. Souza, A.C.S. Appl, Mater. Interfaces **9**, 1609 (2017)
100. H. Jaeger, H. Haas, Science **304**, 78 (2004)
101. D. Sussillo, L. Abbott, Neuron **63**, 544 (2009)
102. G. Tanaka, T. Matsumori, H. Yoshida, K. Aihara, Phys. Rev. Res **4**, L032014 (2022)
103. S. Dubose, J. Endocr. Soc **9**, 60 (2022)

Analog CMOS Implementations of Hardware Neurons for Slow Electronics

Tetsuya Iizuka, Zolboo Byambadorj, and Xiangyu Chen

Abstract To realize CMOS-based hardware spiking neural networks (SNNs), area- and power-efficient neuron circuits are essential. We propose a compact leaky integrate-and-fire (LIF) neuron circuit with a long and tunable time constant, which consists of a capacitor and two pseudo resistors (PRs). The fabricated LIF neuron has a power consumption of 6 μW and a leak time constant of up to 1.2 ms, which is tunable by changing the bias voltage of the PRs. We also propose a neural structure for generating and transmitting time-domain signals, including a neuron module, a synapse module, and two weight modules. The proposed neural structure is driven by a leakage current of MOS transistors and uses an inverter-based comparator to realize a firing function, thus providing higher energy and area efficiency compared to conventional designs. Simulation results of the spiking neural network for reservoir computing with the behavioral model of the proposed neural structure demonstrate the learning function.

1 CMOS-Based Leaky Integrate-and-Fire Neuron with Long and Tunable Time Constant

Deep neural networks (DNNs), which are the second generation of artificial neural networks (ANNs), have been extensively explored in recent years for a growing number of applications. However, their huge energy consumption, especially for the memory access in conventional von Neumann architecture, has forced people to find an alternative way to achieve more power-efficient solutions [1–6]. Spiking neural network (SNN) is one of the attractive solutions as the third generation of

T. Iizuka (✉) · Z. Byambadorj · X. Chen
Systems Design Lab., School of Engineering, The University of Tokyo, Tokyo, Japan
e-mail: iizuka@vdec.u-tokyo.ac.jp

Z. Byambadorj
e-mail: zolboo@silicon.u-tokyo.ac.jp

X. Chen
e-mail: xiangyuchen@silicon.t.u-tokyo.ac.jp

I. H. Inoue (ed.), *Slow Electronics with Reservoir Computing*,
https://doi.org/10.1007/978-981-96-8383-3_4

ANNs that can realize learning functions with low power by mimicking biological neurons. SNNs consist of neurons and synapses and are usually built using a bottom-up approach, which means that each component of the SNNs needs to be designed first [6–12].

Simply speaking, the biological neurons in the brain transmit information by generating spikes [13]. Spiking neural network (SNN), known as the third generation of ANNs [14, 15], play an important role in neuroscience [16, 17] because they can realistically mimic the activity of biological neurons. SNN is an event-based learning that does not require external storage and has been extensively studied as the most promising neuromorphic computing system [18, 19]. Like the human brain, SNN is composed of neurons and synapses [13, 17, 19]. The neuron receives spikes from the synapses (Fig. 1a). If a neuron receives a sufficient number of spikes from several synapses within a certain time window, its membrane potential may reach a critical value, and an action potential is triggered. We say that the neuron has "fired" (Fig. 1b) [13].

The implementation of compact, low-cost SNN hardware systems with real-time online learning capabilities is an ongoing and highly promising research topic [19–22]. A bottom-up approach to build neuromorphic computational systems is commonly used, where we first simulate each component of the system through a single or a set of functional electronic devices. Thus, the first and most important step to implement SNN in hardware is to design its basic building blocks of spiking neurons. Scalability and small area are important consideration [22, 23] in neuronal circuit design, especially because such circuits are intended for very large-scale integrated (VLSI) neural networks consisting of thousands of neurons so that they can mimic the high neuronal density of the human brain.

Among the various neuron models, the leaky integrated-and-fire (LIF) neuron model is simple (usually consisting of resistors and capacitors, as in the inset of Fig. 1a) but has relatively high accuracy [13]. The LIF model has an integration behavior that includes the decay of the membrane potential over time (Fig. 1b). Although

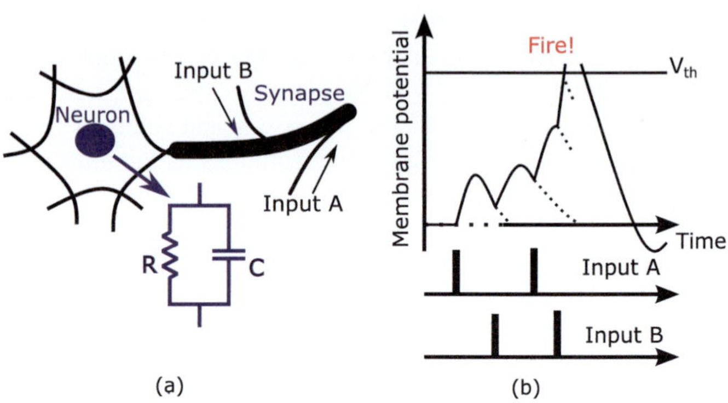

Fig. 1 **a** A simple model of a biological neuron, and **b** changes in neuronal membrane potential with time

it cannot fully reproduce the complex dynamic features of biological neurons, it is suitable for emulating large neuronal networks due to its computational simplicity. The LIF neuron model has become one of the most popular and advantageous neuron models in neuromorphic computing [17]. Given the real-time interaction of hardware neurons with the physical environment, the spike rate of hardware neurons should ideally be similar to that of biological neurons. The time constant of biological neurons is on the order of milliseconds [13, 24, 25], and hence, hardware neurons with considerable R-C time constant need to be designed. A number of artificial neurons with a long time constant have been reported [11, 24, 26–33]. However, these artificial neurons typically use large on-chip resistors and capacitors, occupying a huge chip area and severely hindering system integration [11].

For this reason, we propose a LIF neuron circuit with a long and tunable time constant, using two pseudo-resistors (PRs) [34, 35] consisting of only four transistors. Since the PR can reach a considerable resistance of more than 600 MΩ, even with small on-chip capacitance a long time constant can be achieved, which greatly reduces the area of the chip occupied by each neuron and facilitates the implementation of large-scale neural networks.

1.1 Proposed LIF Neuron Structure Overview

Figure 2 shows the proposed circuit diagram of the integrator-based LIF neuron, consisting of two parts: a leaky-integrator device and a fire device. The leaky-integrator device is essentially a reverse integrator with a leak, and the fire device is simply a comparator.

This leaky integrator device performs a reverse integration when a voltage pulse is applied to the input, with a time constant determined by R_C and C_M. If no pulse signal is injected, the output voltage V_{LI} will leak through the leaky resistor R_L, and eventually reach a steady state. When V_{LI} exceeds the threshold voltage V_{th}, i.e., $V_{LI} < V_{th}$, the comparator will generate a high level, that is, "Fire".

Our design target is to implement a LIF neuron circuit with a realistic biological time constant. However, as mentioned in the previous section, in order to mimic the millisecond time constants of real biological neurons, it is common to use over-sized

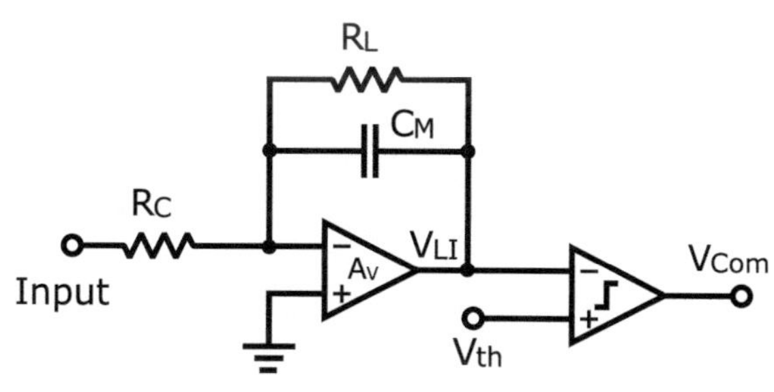

Fig. 2 Circuit diagram of proposed LIF neuron

on-chip capacitors and resistors [29–32], which leads to a significant area occupation (For instance, in [31], a huge capacitor of 10 nF is used.). If we want to achieve a time constant of one millisecond, i.e., the product of the leakage resistance R_L and the membrane capacitance C_M, must be one millisecond. Since the on-chip passive resistor can hardly exceed 100 kΩ with reasonable area, then the capacitance of C_M must be 10 nF, a very large on-chip capacitor. It is almost impossible to integrate thousands of neurons containing such a large capacitor together to form a bio-scale SNN.

To tackle this challenge, we try to use PRs [34], which can achieve large resistance values with a small area. Due to the tunable nature of the PR, it can also be utilized to reset the V_{LI} by turning the MOSFETS in PR on according to the pulse output. To verify the feasibility of this design, we chose the most widely used and low-cost CMOS technology to implement the leaky-integrator device.

1.2 Pseudo-Resistor-Based LIF Neuron Circuit

It is well known that high resistance values can usually be achieved by designing long resistors with multiple folds. This will undoubtedly increase the size cost [36, 37]. In contrast, by utilizing MOSFET transistors that operate in the subthreshold region, high resistances can be achieved without overly occupying the chip area [37]. Since the carrier mobility (hole mobility) of a PMOS device is usually smaller than that (electron mobility) of an NMOS device, the PMOS device exhibits higher resistivity and is widely used for this purpose.

Since the equivalent resistance of the device has exponential dependence on its gate-to-source voltage V_{SG}, the resistance can be tuned in a wide range [34, 35]. However, when its drain-to-source voltage V_{SD} becomes negative, the drain current direction is reversed. In this case, the drain current rapidly increases in the opposite direction as V_{SD} becomes more negative [35], as shown in Fig. 3 (blue line). In order to expand the tuning range and improve linearity, Ref. [34] proposed a PR structure

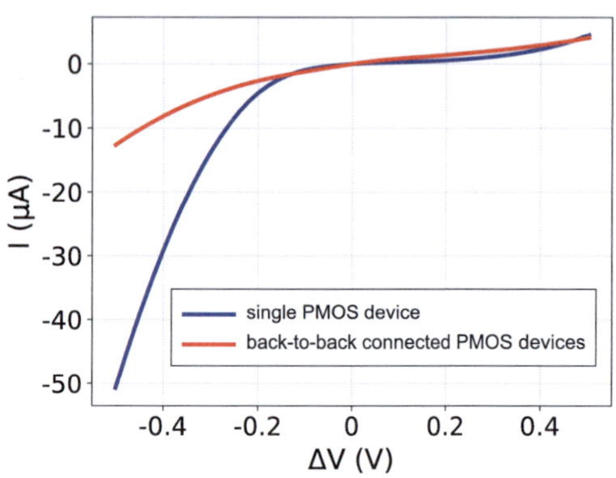

Fig. 3 $I - \Delta V$ characteristics for a single PMOS device (blue line)($V_{SG} = 0.5V$) and a PR composed of two back-to-back connected PMOS devices (red line)($V_{Bias} = 0.5V$)

of two PMOS transistors connected back to back. Figure 4 is a PR designed on the basis of the principle proposed by Ref. [34]. We can control the resistance of the PR through the bias voltage V_{Bias}. As shown in Fig. 3, When the voltage ΔV at both ports is swept from −0.5 V to 0.5 V, the linear high resistance range of the designed PR composed of two back-to-back connected PMOS devices (red line) is nearly doubled compared to a single PMOS device (blue line).

1.3 Chip Fabrication and Measurement Results

The proposed neuron circuit shown in Fig. 5, consisting of two PRs and a 2 pF on-chip capacitor C_M, has been designed, fabricated, and measured with a 65 nm CMOS technology. The die micrograph and layout are shown in Fig. 6. The core area is 54.8 μm × 25.4 μm, and a PR occupies 12.3 μm × 8.1μm. The 2 pF on-chip capacitor is implemented with a metal-insulator-metal (MIM) capacitor, which is placed on top of the circuit. Thus, it does not require a dedicated area.

Figure 7 shows our experimental setup. We place the chip onto the probe station Summit11000 and test it with probes in direct contact with the chip. We use multiple DC voltage sources such as the B2962A to supply power and bias voltages and measure the current, and use the Tektronix AFG31252 arbitrary function generator to provide spikes for the neuron chip. At the same time, we use MSOX6004A Mixed Signal Oscilloscope to observe the output waveform.

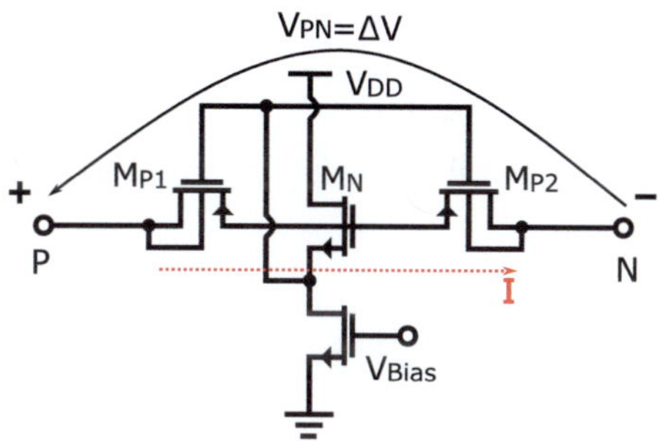

Fig. 4 Designed PR composed of two back-to-back bulk-drain connected PMOS devices

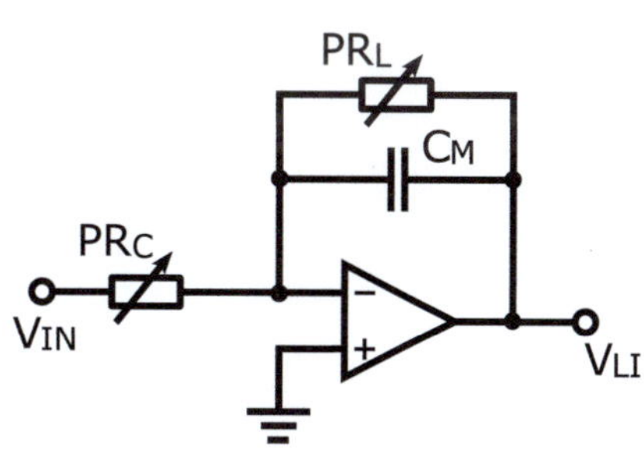

Fig. 5 Proposed leaky integrator neuron circuit utilizing two PRs

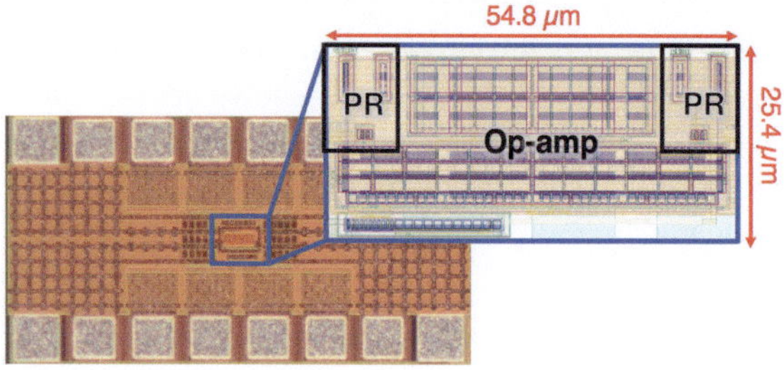

Fig. 6 Chip microphotograph and layout of the proposed neuron circuit implemented in 65 nm CMOS technology

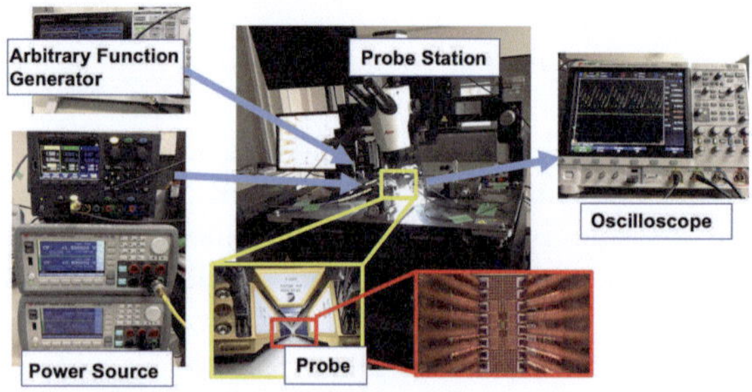

Fig. 7 Experimental setup

Thanks to the use of the PR, one of the advantages of the proposed neuron circuit is the tunability of the time constant. As shown in Fig. 8, while keeping the PR_C constant, the resistance value of PR_L is changed from 130 MΩ (orange line) to 26 MΩ (blue line) by changing the V_{Bias} of PR_L. It is obvious that the upward leakage slope of the 130 MΩ waveform is less than that of 26 MΩ. In other words, the leakage behavior becomes faster with smaller PR_L.

We compared the proposed neuron circuit with other prior neuron circuits as summarized in Table 1. Since some of the prior works do not provide the actual

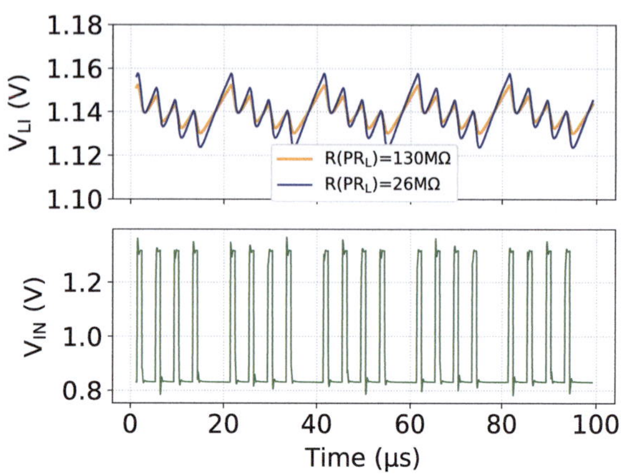

Fig. 8 Tunability of time constant ($PR_L = 130$ MΩ and $PR_L = 26$ MΩ)

Table 1 Comparison with other works

	Ref. [29]	Ref. [31]	Ref. [32]	This work
Process	65 nm CMOS	65 nm CMOS	N/A	65 nm CMOS
Capacitor	0.5 pF	2.36 pF	10 nF	2 pF
Time constant	0.2 μs	19 μs	1 ms	1.2 ms
Power	78.16 μW	15 μW	N/A	6 μW
Core area	538 μm^2	N/A	N/A	1392 μm^2
FOM	40×10^6	8.05×10^6	0.1×10^6	0.6×10^9

values of the power consumption and core area, for the convenience of comparison we define the figure of merit (FOM) that corresponds to the area efficiency as

$$\text{FOM} = \frac{\text{Time constant}}{\text{Capacitance}}. \tag{1}$$

It can be seen that the FOM value of the proposed circuit is much higher than other works. In addition, considering that the occupation area usually increases in proportion to the on-chip capacitor size, compared with Refs. [29, 31], and [32], the proposed LIF neuron circuit can realize the same order of time constant with about 1500×, 74×, 6000× smaller area, respectively. The power consumption of the proposed circuit is much lower than Refs. [29, 31].

1.4 Summary

We have designed a leaky integrate and fire neuron circuit with long and tunable time constants using a 65 nm CMOS process and realized it on a small chip area. Through the measurement, even in the case of process variation, PR can reach close to 600 MΩ, that is to say, a time constant of 1.2 ms is achieved in an area of only 54.8 μm × 25.4 μm. Compared with other neuron circuits that have been reported, the proposed circuit can achieve the same order of time constant in a small area, which is advantageous in realizing a large-scale SNN. The tunability of the R-C time constant in the proposed circuit is particularly useful for emulating the behavior of emerging neuronal devices.

2 CMOS-Based Time-Domain Analog Neuron and Synapse Circuits

Many hardware implementations of pulsed neurons or synapses have been reported [38–46]. To implement the leaky integrate function of neurons, conventional designs including our previous work [8, 11] usually build integrators with operational amplifiers (op-amps) [39] and often use large on-chip capacitors and resistors to mimic the millisecond time constants of biological neurons [41, 42]. Moreover, to implement the neuron "fire" function, a dedicated circuit structure of a continuous-time or a clocked comparator is usually used to set the threshold for neuron excitation [38–41, 45]. The bias current of the continuous-time comparator undoubtedly increases the power consumption of the neuron, while the clocked comparator requires additional clock signal distribution and the complex comparator structure occupies a large chip area. While more advanced processes can achieve low power consumption by reducing supply voltage and static leakage current [46], this also leads to a narrower dynamic range, smaller available margin, and degraded noise immunity of the voltage/current-domain analog circuits [47]. This is detrimental to conventional neural networks that use analog quantities such as voltage and current to communicate with each other. On the other hand, thanks to the scaled transistors that have an improved operation speed with sharp signal transitions, the analog information can be represented more efficiently in the time domain, i.e. a time interval of two signal transitions. This so-called time-domain circuit has another advantage in its power efficiency as it often consists of inverters or logic gates that ideally consume no DC power [47, 48]. Thus, time-domain circuits are ideal for future implementations of low-power SNNs.

We propose a novel neural structure for generating and transmitting time-domain signals to compose a time-domain neural network [49]. The integrated structure includes neuron and synapse modules that respectively generate and transmit time-domain signals, as well as weight modules for learning functions. One of our main target applications is reservoir computing, which processes information related to human activity. Our application targets simpler and less data-intensive processing such as bio-signals. In reservoir computing, learning functions such as ECG and speaker recognition, as well as handwriting recognition, can be implemented using only a few hundreds of neurons. Ref. [50] shows that learning performance improves when the time constants of the input effects are matched between the target function and reservoir dynamics, and we use millisecond time constants as a design target for a neural structure that will be used to process time-series information of human activities. We use the behavioral model of the proposed neural structure to construct the SNN for reservoir computing and implement the learning function, which proves that our proposed neural structure can be used for reservoir computing.

2.1 Proposed Time-Domain Neural Structure

The designed and fabricated neural structure is shown in Fig. 9a, which is based on the proposed neuron, synapse, and weight modules, which will be described in detail below. In this structure, the input of the neuron module is connected to two weight modules, one for tuning the inhibitory signal and the other for the excitatory signal. We fabricated the proposed neural structure shown in Fig. 9a with a 65 nm CMOS technology. The micrograph of the chip is shown in Fig. 9b, where the die area of neuron, synapse, and weight modules are 127 μm^2, 231 μm^2 and 525 μm^2, respectively.

The LIF neuron model consists mainly of a membrane capacitor, a leaky resistor, and a voltage comparator. Neurons receive signals from other neurons via synapses, and the soma generates action potentials in response to these external signals. If a neuron receives a sufficient number of spikes through the synapse, its membrane potential will reach a threshold value, causing the neuron to "fire" [8, 51, 52].

The use of inverters to implement the "fire" function is already known as an alternative to comparators. Ref. [53] has proposed an inverter-based neuron, which is well suited for use in the proposed neural structure, and therefore the neuron used in this study was designed based on Ref. [53], which is shown in Fig. 10a. It consists of an input device, a leaky integrator device, a fire device, and a delay device. Originally in Ref. [53], the circuit is not assumed to be designed as an element to build a neural network and thus does not have a structure to receive excitatory and inhibitory signals. In the proposed circuit, on the other hand, the input device that consists of M_1 and M_2 receives an excitatory input and an inhibitory input, respectively. The inputs to M_1 and M_2 are narrow pulse signals as shown in Fig. 10a, which is generated from a pre-stage synapse. The activity of the pre-stage synapse is represented by the pulse frequency, and the coupling weight is represented by the pulse width. When more than one pre-stage synapses are connected to compose a network, the multiple pulses

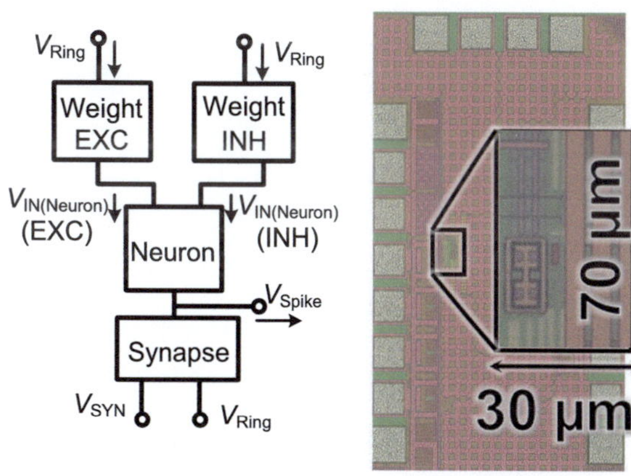

(a) (b)

Fig. 9 **a** The proposed structure and **b** a micrograph of the chip

can be applied through OR logic, or by adding input devices connected in parallel. With parallel input devices, the neuron circuit can accept multiple pulses even at the same time.

In a leaky-integrator device, C_{mem} represents the cell membrane of the neuron, and M_5 can be regarded as a leaky resistor in the resting state. When there is no external input to the input device, the capacitor is charged by the leakage current of M_3 and M_4, and the membrane potential V_{mem} increases continuously with the inflow of the leakage current (the current is integrated as shown in Fig. 10b(i)). At this point, since M_5 is in the off state, it can be considered as a resistor in parallel with the capacitor, i.e. leaky resistor, capable of achieving a long time constant.

Once V_{mem} rises to the threshold voltage $V_{th(Fire)}$, the fire device is activated (Fig. 10b(ii)). In conventional designs, LIF neurons mostly use dedicated circuit structures of a continuous-time or a clocked comparator to set the threshold voltage. This is not friendly for building SNNs that should be energy efficient and bio-scale, like the brain. In this study, the fire device is implemented by an inverter-based comparator that can set the threshold voltage by two transistors instead of a continuous-time or clocked comparator. To realize an accurate threshold voltage for

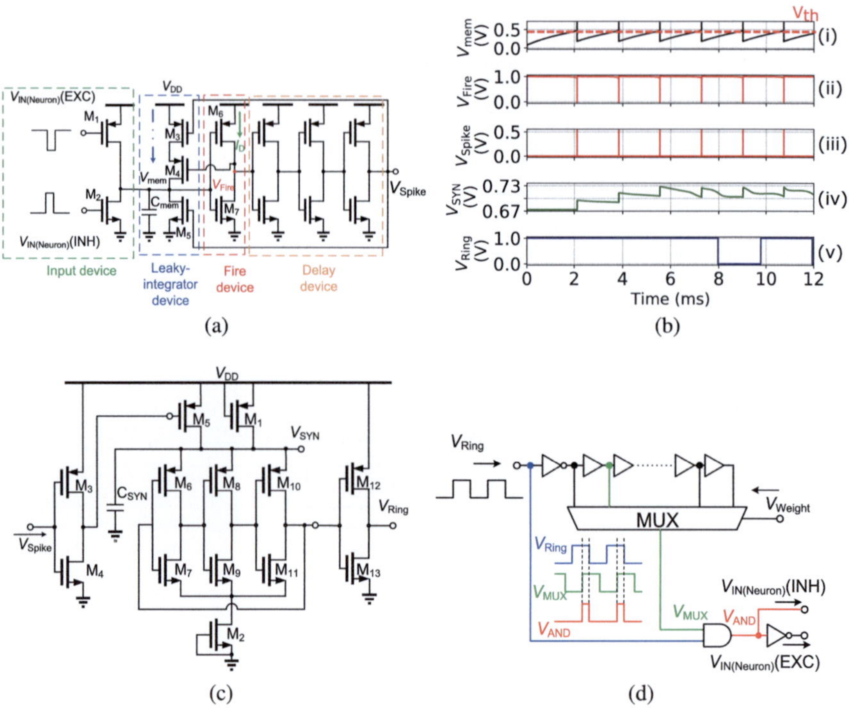

Fig. 10 **a** Circuit diagram of the proposed neuron module, **b** behaviors of proposed LIF neuron and synapse modules, **c** circuit diagram of the proposed synapse module, and **d** circuit diagram of the proposed weight module

an inverter-based comparator, we may use an auto-zeroing technique that periodically senses, stores, and cancels the offset with switches and capacitors [54]. However, it requires multiple-phase clocks to control the switches, thus is not suitable for area-and-power-efficient reservoir implementations. Though with a simple inverter-based comparator, there may be a threshold variation due to process, voltage, and temperature fluctuations, it can be seen as mimicking the difference between individuals of real neurons. In addition, the learning function is able to compensate for threshold differences and process variations [55]. When there is an excitatory pulse input, M_1 will be turned on instantaneously, which causes more current to charge C_{mem} and V_{mem} to rise rapidly. Conversely, an inhibitory pulse input signal will cause M_2 to turn on momentarily, causing C_{mem} to charge slower or even discharge through M_2, which in turn slows down the rate of V_{mem} rise or makes it fall.

When the fire device is activated, it generates a low level of V_{Fire} to be connected to M_4, which will increase the current to charge the membrane capacitor C_{mem}, resulting in an instantaneous increase of the membrane potential V_{mem}, which promotes the triggering of the fire device. This mimics the influx of Na^+ into the cell membrane prompting a rapid increase in membrane voltage, i.e., a positive feedback effect. Finally, the low level of V_{Fire} generated from the fire device is converted to a high level of V_{Spike} (Fig. 10b(iii)) by a delay device that includes a three-stage inverter and connects the V_{Spike} to M_3 and M_5, resetting V_{mem} to zero. This process mimics the activation of K^+ channels in biological neurons, resulting in the outward flow of K^+ ions and the eventual return of the cell membrane to its resting state.

Synapses are essential modules in SNNs, as neurons are interconnected by them. We have designed a neuron module for generating time-domain signals, and then we need a transmission medium, i.e., a synapse, to transmit this time-domain signal to other neurons. To compose a complete neural network, we design a synapse module based on frequency signals, as shown in Fig. 10c. The synapse consists mainly of a voltage-controlled ring oscillator operating under a leakage current, which is composed of a three-stage inverter (M_6, M_7, M_8, M_9, M_{10} and M_{11}). The previous neuron circuit fires and generates a spike V_{Spike}, which is inverted by an inverter, making M_5 open for a short time, and the current flowing through M_5 charges C_{SYN}, which will increase V_{SYN}. Once V_{SYN} reaches the voltage that triggers the oscillation, the ring oscillator begins to oscillate (Figs. 10b(iv) and (v)). If the preceding neuron does not fire for a long time, V_{SYN} will leak until the initial state, at which point the synapse becomes inactive again. Since V_{SYN} is equivalent to the supply voltage of the ring oscillator, the current flowing out of M_5 controls V_{SYN} and thus the frequency of the ring oscillator.

SNNs achieve learning function by adjusting the weights; therefore, we propose a weight module that is compatible with the proposed time-domain neuron and synapse modules described above, as shown in Fig. 10d. The proposed weight module tunes the time-domain information, which is the width of the output pulses. This module consists of a delay line, a multiplexer, and an AND gate. V_{Ring} is the square wave signal from the synapse that will pass through the delay line. V_{Weight} is the digital code that represents weight, which is determined after learning and is used to control the multiplexer. The width of the output pulse that corresponds to the time-domain weight is adjusted according to which tap in the inverter chain is selected by the multiplexer. As mentioned earlier, if the excitatory or inhibitory pulse width is wide, the voltage V_{mem} in the subsequent neuron is charged or discharged faster, respectively. This corresponds to a large weight. In this study, we chose a multiplexer with 16 inputs, i.e. four-bit weights (0000 to 1111). The output of the weight module is connected to the input device of the subsequent neuron circuits. The frequency of the pulse (pulse spacing) and the width of the pulse act simultaneously on the neuron to change its activity. The frequency of the pulse is determined by the output frequency of the previous synapse, while the coupling strength depends on the width of the pulse output determined by the weight module.

2.2 Measurement Results

Figure 11a shows the experimental setup used to test the fabricated neural structure chip (Fig. 9b), where the chip was placed on a probe station Summit11000 and tested with probes in direct contact with it. In the experiments, we assume that the inputs of the two weight modules are the pre-stage synapses, which are emulated by the arbitrary function generators. The output of the neuron is connected to the synapse module, and the output of which will be varied in response to the change in the output of the neuron. We used a Tektronix AFG31252 arbitrary function generator as a pre-stage synapse to provide square wave signals for our fabricated neural circuits. At the same time, we observed the output waveforms using oscilloscopes (Keysight MSOX6004A and DSOX93304Q).

The experimental results are shown in Figs. 11b–d. To verify the effect of weights on the firing rate of neurons, we fixed the frequency of the pre-stage synapse output (function generator) 100 Hz and observed the change in the firing rate of neurons for 4 chips by adjusting the weights module. We averaged the spike frequencies 1024 times over a time range of 100 ms to derive the corresponding neuronal firing frequency under each weight setting, as shown in Fig. 11b. The proposed neuron is basically firing with the rate determined by the leakage currents into and out from C_{mem} in balance and input from the previous stage modulates it. We can see that when the weights become larger, the neuron module firing frequency becomes larger. Mainly due to the process variation of the FETs, the firing frequency fluctuates in about $\pm$10–17 % over 4 chips. Especially for the use in a reservoir, however, due to the random weights in its recurrent connections, these random variations should be compensated for during the learning process in the output weights.

Figure 11c compares the variation of neuron fire times depending on the signal from the pre-stage synapse. The insets (i), (ii) and (iii) of Fig. 11c show the cases 100 Hz inhibitory input (weight is set to 1100), no input, and 100 Hz excitatory input (weight is set to 1100), respectively, from which we can see that the inhibitory input decreases the fire frequency of the neuron and increases the fire interval, while the excitatory input works as the opposite of the inhibitory input. The experimental results show that the firing interval of the proposed neuron is on the order of milliseconds,

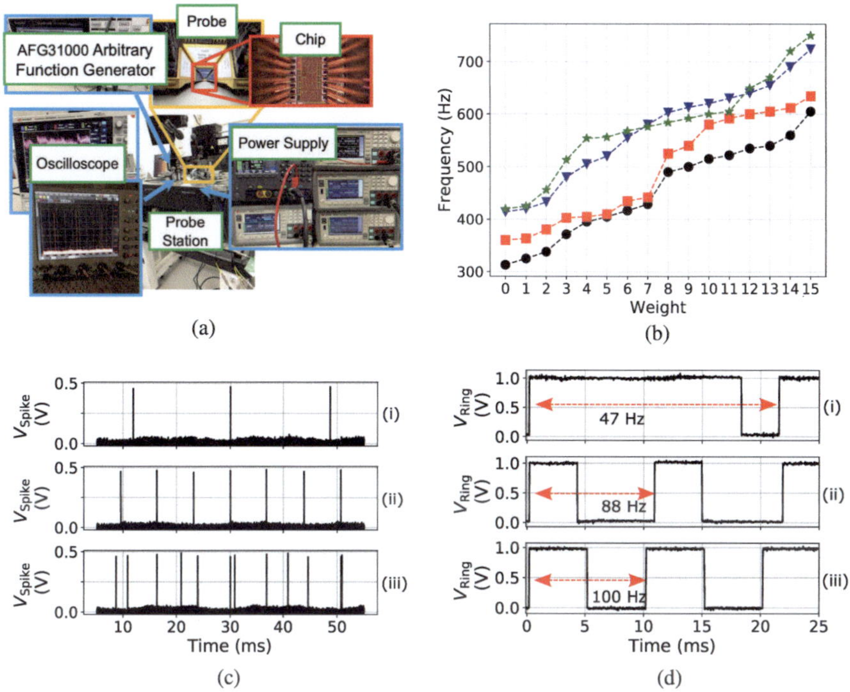

Fig. 11 **a** A photo of the experimental setup, **b** the measured firing rate of the neuron for 4 chips, **c** the measured waveforms of the neuron output, and **d** the measured waveforms of the synapse output

which is in accordance with the feature of biological neurons having millisecond time constants. When no signal is fed from the pre-stage synapse, the power consumption is about 800 pW, generating about 20 spikes in a 100 ms cycle. From this, it can be roughly estimated that each spike consumes about 4 pJ of energy. Subsequently, the insets (i), (ii), and (iii) of Fig. 11c were used as input signals to the synapse to influence V_{Ring}. The measured V_{Ring} waveforms in these three cases are shown in Fig. 11d. The average of the frequencies for each case measured in a 5 s time period are 41 Hz, 90 Hz and 98 Hz, respectively. The feasibility of this synapse output frequency range will be validated with system-level simulations in the following discussion.

Table 2 shows the performance comparison among stand-alone neuron circuits. The proposed neuron circuit has advantages in terms of energy consumption and area occupation. The designs in Refs. [38–41] used a continuous-time or a clocked comparator, and these designs take up a large amount of chip area as well as power consumption. The neuron fabricated in a non-CMOS process proposed in Ref. [43] does not require a comparator, which leads to an advantage in the area. However, its energy consumption is relatively high and these particular technologies are less mature and thus more costly compared to standard CMOS processes. Both Refs. [44, 46] are

Table 2 Performance comparison of stand-alone neuron circuits

Ref.	Technology	Energy per spike (pJ)	Single neuron area (μm^2)	Frequency (Hz)	Neuron model	Sim. or Meas.
[38]	350 nm CMOS	900	2573	100	IF	Meas.
[39]	180 nm CMOS	9.3	10^4	N/A	LIF	Meas.
[40]	65 nm CMOS	41	538	300	LIF	Sim.
[41]	65 nm CMOS	200	3363	1.9×10^6	AdEx-IF[1]	Meas.
[43]	32 nm SOI MOSFET	35	1.8	10^6	LIF	Meas.
[44]	22 nm FD-SOI	14	900	30	AdEx-IF[1]	Meas.
[46]	22 nm CMOS	1.17/0.36	70	30/100	LIF	Sim.
This work	65 nm CMOS	4	127	230	LIF	Meas.

1, Adaptive-exponential integrate-and-fire

being fabricated in an advanced process. However, compared to this work, Ref. [44] does not have an advantage in terms of energy consumption and area. Although Ref. [46] shows better energy efficiency with simulation results, when normalized by the technology node, the proposed neuron achieves better area efficiency.

To demonstrate the feasibility of the proposed spiking neuron and the ring oscillator-based synapse circuits, a behavioral simulation is carried out in MATLAB environment as shown in Fig. 12a. In this simulation, 100 neurons are used with random recurrent connections with the proposed synapse and weighting modules. The proposed weight modules are applied only in the reservoir layer and their weights are assigned randomly in advance and fixed during the learning process. To establish a realistic simulation, the output frequency range of each synapse is set from 15 Hz to 200 Hz based on the actual measurement results. The recursive least square (RLS) algorithm is used to train the output weights as introduced in Ref. [56]. A 10 Hz sinewave, which corresponds to the time-scale of human-activity-related information, is used as an example of a supervisory input signal. The supervisory and the trained output signals are shown in Fig. 12b(i). The feedback signal from the output is converted to excitatory and inhibitory pulse trains whose frequencies are in proportion to the absolute value of the output amplitude as shown in Figs. 12b(ii) and (iii), respectively. After 5 periods of supervisory signal, the output weights are fixed and the SNN generates the learned signal by itself, which demonstrates the feasibility of the proposed neural structures for learning function. Figure 12b(iv) shows the learning results where the neuron firing rates and the synapse output frequencies randomly fluctuated from the nominal values with the standard deviation of $\pm 15\%$ to

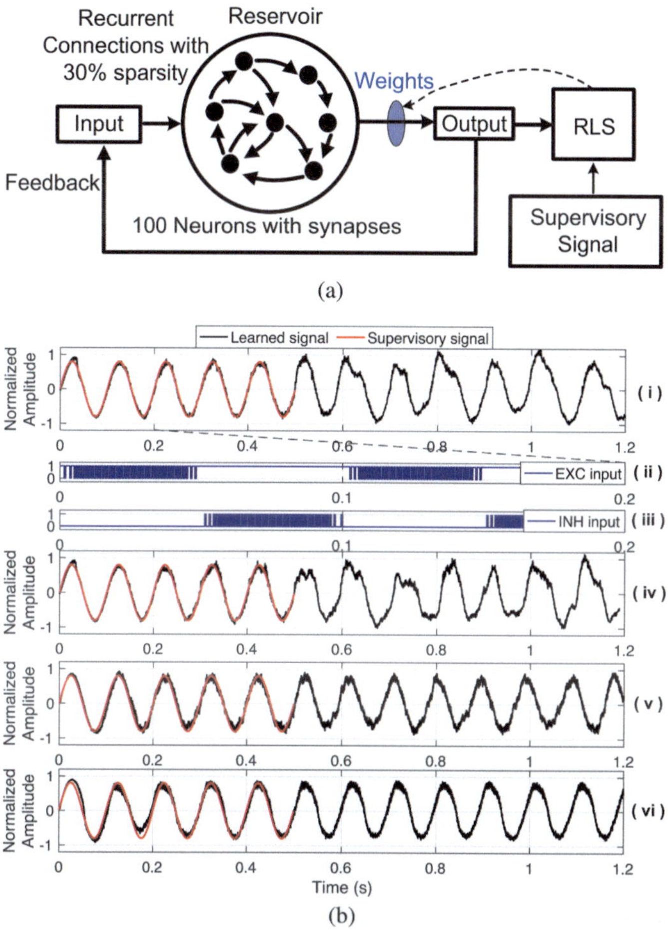

Fig. 12 **a** The behavioral model of the SNN for reservoir computing based on the proposed neural structure. **b** The system-level behavioral simulation results: (i) based on a model with 15 Hz–200 Hz frequency tuning range, zoomed-in view of the (ii) excitatory and (iii) inhibitory input signals converted from the output, (iv) based on a model with process variation in neurons and synapses, (v) based on 15 Hz–2 kHz and (vi) 15 Hz–20 kHz frequency tuning ranges

consider a more realistic situation, which demonstrates the learning capability even with process variation. We have also found from these simulations that to further improve the learning capability, the output frequency tuning range of the synapse should be increased, which can be done by optimizing the synapse circuit. For example, with the extended frequency tuning ranges 15 Hz–2 kHz and 15 Hz–20 kHz, the learned signals become smoother to better reproduce the supervisory signal as shown in Figs. 12b(v) and (vi), respectively.

2.3 Summary

We have proposed a neural structure for generating and transmitting time-domain signals. The proposed neuron and synapse occupy areas of 127 μm^2 and 231 μm^2, respectively. This structure does not use op-amps and continuous-time or clocked comparators, while the firing function is realized with an inverter-based comparator to provide advantages in area and power consumption. The proposed time-domain neural structure benefits from scaled process technologies compared to conventional voltage/current-domain designs. Actual chip fabrication and measurement results demonstrate the temporal signal communication function with millisecond time constants. The proposed time-domain neural structure is well suited for building spiking neural networks for processing real-time time-series information for human-computer interaction.

Acknowledgements This work is supported by Japan Science and Technology Agency (JST) CREST Grant Number JPMJCR19K2.

References

1. Y. Zhang, P. Qu, Y. Ji, W. Zhang, G. Gao, G. Wang, S. Song, G. Li, W. Chen, W. Zheng, F. Chen, J. Pei, R. Zhao, M. Zhao, L. Shi, Nature **586**, 378 (2020)
2. D. Shin, H.J. Yoo, Proceedings of the IEEE **108**, 1245 (2020)
3. Y. LeCun, Y. Bengio, G. Hinton, Nature **521**, 436 (2015)
4. T. Kohno, K. Aihara, AIP Conference Proceedings **1028**(1), 113 (2008). https://doi.org/10.1063/1.2965080
5. E. Chicca, G. Indiveri, Appl. Phys. Lett. **116**(12), 120501 (2020). https://doi.org/10.1063/1.5142089
6. Y. Bo, P. Zhang, Y. Zhang, J. Song, S. Li, X. Liu, J. Appl. Phys. **127**(24), 245101 (2020). https://doi.org/10.1063/5.0004139
7. K. Yang, A. Sengupta, Appl. Phys. Lett. **116**(4), 043701 (2020). https://doi.org/10.1063/1.5138951
8. X. Chen, T. Yajima, I.H. Inoue, T. Iizuka, Jpn. J. Appl. Phys. **61**, SC1051 (2022)
9. W. Maass, Neural. Netw. **10**, 1659 (1997)
10. S.S. Radhakrishnan, A. Sebastian, A. Oberoi, S. Das, S. Das, Nat. Commun. **12**, 2143 (2021)
11. X. Chen, T. Yajima, I.H. Inoue, T. Iizuka, Extended Abstract of International Conference on Solid State Devices and Materials (SSDM) pp. 682–683 (2021)
12. D.S. Jeong, J. Appl. Phys. **124**(15), 152002 (2018). https://doi.org/10.1063/1.5042243
13. W. Gerstner, W.M. Kistler, R. Naud, L. Paninski, *Neuronal Dynamics: From Single Neurons To Networks And Models Of Cognition* (Cambridge University Press, New York, 2014)
14. W. Maass, Neural. Netw. **10**(9) (1997)
15. D. Chatterjee, A. Kottantharayil, IEEE Electron Device Letters **40**(8), 1301 (2019). https://doi.org/10.1109/LED.2019.2924259
16. J. Schemmel, D. Bruderle, A. Grubl, M. Hock, K. Meier, S. Millner, in *2010 IEEE International Symposium on Circuits and Systems (ISCAS)* (2010), pp. 1947–1950. https://doi.org/10.1109/ISCAS.2010.5536970
17. J.Q. Yang, R. Wang, Y. Ren, J.Y. Mao, Z.P. Wang, Y. Zhou, S.T. Han, Adv. Mater. **32**(2003610) (2020)

18. P.A. Merolla, J.V. Arthur, R. Alvarez-Icaza, A.S. Cassidy, J. Sawada, F. Akopyan, B.L. Jackson, N. Imam, C. Guo, Y. Nakamura, B. Brezzo, I. Vo, S.K. Esser, R. Appuswamy, B. Taba, A. Amir, M.D. Flickner, W.P. Risk, R. Manohar, D.S. Modha, Science **345**(668) (2014)
19. G. Indiveri, B. Linares-Barranco, T.J. Hamilton, A. van Schaik, R. Etienne-Cummings, T. Delbruck, S.C. Liu, P. Dudek, P. Hafliger, S. Renaud, J. Schemmel, G. Cauwenberghs, J. Arthur, K. Hynna, F. Folowosele, S. Saighi, T. Serrano-Gotarredona, J. Wijekoon, Y. Wang, K. Boahen, Frontiers in Neuroscience **5** (2011). https://doi.org/10.3389/fnins.2011.00073. https://www.frontiersin.org/journals/neuroscience/articles/10.3389/fnins.2011.00073
20. G.K. Chen, R. Kumar, H.E. Sumbul, P.C. Knag, R.K. Krishnamurthy, IEEE Journal of Solid-State Circuits **54**(4), 992 (2019). https://doi.org/10.1109/JSSC.2018.2884901
21. N. Qiao, H. Mostafa, F. Corradi, M. Osswald, F. Stefanini, D. Sumislawska, G. Indiveri, Frontiers in Neuroscience **9** (2015). https://doi.org/10.3389/fnins.2015.00141. https://www.frontiersin.org/journals/neuroscience/articles/10.3389/fnins.2015.00141
22. J.s. Seo, B. Brezzo, Y. Liu, B.D. Parker, S.K. Esser, R.K. Montoye, B. Rajendran, J.A. Tierno, L. Chang, D.S. Modha, D.J. Friedman, in *2011 IEEE Custom Integrated Circuits Conference (CICC)* (2011), pp. 1–4. https://doi.org/10.1109/CICC.2011.6055293
23. J.H.B. Wijekoon, P. Dudek, Neural Netw. **21**(524) (2008)
24. S.C. Liu, R. Douglas, IEEE Transactions on Neural Networks **15**(5), 1305 (2004). https://doi.org/10.1109/TNN.2004.832725
25. M.A. Nahmias, B.J. Shastri, A.N. Tait, P.R. Prucnal, IEEE Journal of Selected Topics in Quantum Electronics **19**(5), 1 (2013). https://doi.org/10.1109/JSTQE.2013.2257700
26. J.J. Wang, Q. Yu, S.G. Hu, Y. Liu, R. Guo, T.P. Chen, Y. Yin, Y. Liu, Appl. Phys. Lett. **115**(243701) (2019)
27. S. Moradi, S.A. Bhave, R. Manohar, in *2017 IEEE Symposium Series on Computational Intelligence (SSCI)* (2017), pp. 1–5. https://doi.org/10.1109/SSCI.2017.8285437
28. A. Joubert, B. Belhadj, R. Héliot, in *2011 IEEE 9th International New Circuits and systems conference* (2011), pp. 9–12. https://doi.org/10.1109/NEWCAS.2011.5981206
29. A. Joubert, B. Belhadj, O. Temam, R. Heliot, in *The 2012 International Joint Conference on Neural Networks (IJCNN)* (2012), pp. 1–5. https://doi.org/10.1109/IJCNN.2012.6252600
30. A. Velichko, P. Boriskov, IEEE Transactions on Circuits and Systems II: Express Briefs **67**(12), 3477 (2020). https://doi.org/10.1109/TCSII.2020.2997117
31. S.A. Aamir, P. Muller, A. Hartel, J. Schemmel, K. Meier, in *ESSCIRC Conference 2016: 42nd European Solid-State Circuits Conference* (2016), pp. 71–74. https://doi.org/10.1109/ESSCIRC.2016.7598245
32. M.J. Rozenberg, O. Schneegans, P. Stoliar, Sci. Rep. **9**(11123) (2019)
33. G. Indiveri, E. Chicca, R. Douglas, IEEE Transactions on Neural Networks **17**(1), 211 (2006). https://doi.org/10.1109/TNN.2005.860850
34. A. Tajalli, Y. Leblebici, E. Brauer, Electron. Lett. **44**(349) (2008)
35. A. Tajalli, Y. Leblebici, *Extreme Low-Power Mixed Signal IC Design: Subthreshold Source-Coupled Circuits* (Springer, New York, 2010)
36. E. Guglielmi, F. Toso, F. Zanetto, G. Sciortino, A. Mesri, M. Sampietro, G. Ferrari, IEEE Journal of Solid-State Circuits **55**(8), 2094 (2020). https://doi.org/10.1109/JSSC.2020.2973639
37. R. Puddu, C. Carboni, L. Bisoni, G. Barabino, D. Pani, L. Raffo, M. Barbaro, IEEE Transactions on Circuits and Systems II: Express Briefs **64**(7), 762 (2017). https://doi.org/10.1109/TCSII.2016.2603533
38. G. Indiveri, E. Chicca, R. Douglas, IEEE Trans. Neural. Netw. **17**, 211 (2006)
39. X. Wu, V. Saxena, K. Zhu, S. Balagopal, I.E.E.E. Trans, Circuits Syst. II, Express Briefs **62**, 1088 (2015)
40. A. Joubert, B. Belhadj, O. Temam, R. Heliot, The 2012 International Joint Conference on Neural Networks (IJCNN) pp. 1–5 (2012)
41. S.A. Aamir, P. Muller, G. Kiene, L. Kriener, Y. Stradmann, A. Grubl, J. Schemmel, K. Meier, IEEE Transactions on Biomedical Circuits and Systems **12**(5), 1027 (2018). https://doi.org/10.1109/TBCAS.2018.2848203
42. A. Basu, P.E. Hasler, IEEE Trans. Circuits Syst. I, Reg. Papers **57**(11), 2938 (2010)

43. S. Dutta, V. Kumar, A. Shukla, N.R. Mohapatra, U. Ganguly, Scientific Reports **7**, 8257 (2017)
44. A. Rubino, M. Payvand, G. Indiveri, 2019 26th IEEE International Conference on Electronics, Circuits and Systems (ICECS) pp. 458–461 (2019)
45. S.A. Aamir, Y. Stradmann, P. Muller, C. Pehle, A. Hartel, A. Grubl, J. Schemmel, K. Meier, IEEE Trans. Circuits Syst. I, Reg. Papers **65**, 4299 (2018)
46. R.M. Saber Moradi, Sunil A. Bhave, 2017 IEEE Symposium Series on Computational Intelligence (SSCI) pp. 1–5 (2017)
47. K. Asada, T. Nakura, T. Iizuka, M. Ikeda, IEICE Electronics Express **15**(6), 20182001 (2018)
48. R. Staszewski, K. Muhammad, D. Leipold, C.M. Hung, Y.C. Ho, J. Wallberg, C. Fernando, K. Maggio, R. Staszewski, T. Jung, J. Koh, S. John, I.Y. Deng, V. Sarda, O. Moreira-Tamayo, V. Mayega, R. Katz, O. Friedman, O. Eliezer, E. de Obaldia, P. Balsara, IEEE J. Solid-State Circuits **39**(12), 2278 (2004)
49. X. Chen, Z. Byambadorj, T. Yajima, H. Inoue, I.H. Inoue, T. Iizuka, Applied Physics Letters **122**(7), 074102 (2023). https://doi.org/10.1063/5.0136627
50. C. Gallicchio, A. Micheli, Neural. Netw. **24**(5), 440 (2011)
51. L.F. Abbott, P. Dayan, *Theoretical Neuroscience* (The MIT Press, 2005)
52. W. Gerstner, W.M. Kistler, *Spiking Neuron Models: Single Neurons, Populations* (Cambridge University Press, 2012)
53. T. Yajima, Sci. Rep. **12**, 1150 (2022)
54. B. Razavi, *Principle of Data Conversion System Design* (Wiley-IEEE Press, 1995)
55. T. Wunderlich, A.F. Kungl, E. Muller, A. Hartel, Y. Stradmann, S.A. Aamir, A. Grubl, A. Heimbrecht, K. Schreiber, D. Stockel, C. Pehle, S. Billaudelle, G. Kiene, C. Mauch, J. Schemmel, K. Meier, M.A. Petrovici, Front. Neurosci. **13**(260), 1 (2019)
56. D. Sussillo, L. Abbott, Neuron **63**(4), 544 (2009). https://doi.org/10.1016/j.neuron.2009.07.018. https://www.sciencedirect.com/science/article/pii/S0896627309005479

Learning and Inference in Slow Electronics: Numerical Simulation

Hiroto Tamura, Hisashi Inoue, and Takeaki Yajima

Abstract This chapter explores applications of slow electronics for temporal pattern recognition through numerical simulations of reservoir computing models. We demonstrate how the inherent time constants of slow electronic systems can be advantageously matched to tasks with similar temporal dynamics. Two primary applications are presented: handwriting authentication and blood glucose prediction. For handwriting authentication, we implement time series anomaly detection using both leaky integrator and leaky integrate-and-fire (LIF) neuron models, showing that reservoirs with longer time constants more effectively capture the unique temporal dynamics of individual handwriting patterns. For blood glucose prediction, we simulate a system using protonic devices as short-term memory elements, successfully forecasting glucose levels 10–30 min in advance–a critical timeframe for preventive healthcare interventions. The prediction accuracy remains comparable to conventional software implementations while offering the potential for ultra-low power operation. These simulations establish an important foundation for implementing slow electronics-based computing systems that effectively process human-scale temporal data while maintaining high performance within ultra-low power constraints.

H. Tamura (✉)
Graduate Schools for Law and Politics, and International Research Center for Neurointelligence (IRCN), The University of Tokyo, Tokyo, Japan
e-mail: h-tamura@g.ecc.u-tokyo.ac.jp

H. Inoue
National Institute of Advanced Industrial Science and Technology (AIST), Tsukuba, Japan
e-mail: hisashi.inoue@aist.go.jp

T. Yajima
Department of Electrical Engineering and Computer Science, Kyushu University, Fukuoka, Japan
e-mail: yajima@ed.kyushu-u.ac.jp

I. H. Inoue (ed.), *Slow Electronics with Reservoir Computing*,
https://doi.org/10.1007/978-981-96-8383-3_5

1 Introduction

This chapter explores practical applications of slow electronics through the lens of reservoir computing models. Reservoir computing provides an ideal framework for implementing slow electronics due to its fixed recurrent connection weights (see Chap. 2 for details). This approach is particularly well-suited for processing time-series data with longer time constants, aligning with the operational characteristics of slow electronic systems.

We present numerical simulations of specific application tasks that leverage the advantageous properties of slow electronics. In particular, we focus on tasks with relatively slow temporal dynamics that match the inherent time constants of our proposed electronic systems: handwriting authentication and blood glucose prediction.

In Sects. 2–3, the handwriting authentication system demonstrates how slow electronics can capture the unique temporal patterns in human motor behavior by applying time series anomaly detection techniques. This application showcases how the temporal integration properties of slow electronic systems can be harnessed for biometric security applications, where the system must recognize legitimate variations in a person's handwriting while rejecting unauthorized attempts. For simplicity and to establish a theoretical foundation, we conduct simulations using simple neuron models–namely the leaky integrator model (Sect. 2) and the leaky integrate-and-fire (LIF) model (Sect. 3)–instead of actual slow electronics devices.

In Sect. 4, the blood glucose prediction application illustrates how slow electronics can contribute to preventive healthcare by forecasting glucose levels 10–30 min in advance. Using reservoir computing with simulated protonic devices as short-term memory elements, our system processes continuous glucose monitoring data to provide timely predictions that could potentially help in medication management and disease prevention.

These simulations serve as an important stepping stone towards the implementation of actual slow electronics-based reservoir computing systems, demonstrating that even with ultra-low power hardware constraints, such systems can achieve high performance for slow time-scale data.

2 Handwriting Authentication Using Anomaly Detection

2.1 Anomaly Detection in Time Series Data

Time series anomaly detection [1, 2] constitutes a fundamental challenge within data mining research. Whereas offline techniques examine retrospective datasets, online detection methods identify anomalies in current data using only historical observations [3]. The online capability is vital across domains including industrial control, infrastructure monitoring, and healthcare systems.

State-of-the-art deep learning methods [4–6] achieve notable success in anomaly detection due to their capacity to model complex temporal dependencies. However, these approaches typically need substantial computational cost during training, which becomes especially problematic in dynamic environments where steady-state conditions shift relatively quickly, rendering models obsolete before they can be fully trained. Furthermore, deep learning requires large amounts of training data, making it challenging to implement in scenarios when historical data is limited.

Reservoir computing (RC) [7–9] addresses these challenges through a fixed, randomly initialized recurrent neural network that transforms temporal inputs into high-dimensional state representations. The fixed internal connections of the reservoir require training only of the output layer, substantially reducing computational costs during the learning phase. Another key advantage of RC is its ability to function effectively with significantly less training data compared to deep learning methods. These properties make RC particularly suitable for anomaly detection in slow electronics contexts where both data efficiency and computational efficiency are critical considerations.

2.2 *Reservoir Computing for Time Series Anomaly Detection*

RC provides an efficient framework for processing temporal data through a unique neural architecture. As shown in Fig. 1, an RC system contains two main elements: a reservoir with recurrently connected neurons that transform inputs into high-dimensional state representations, and a readout layer that extracts patterns from these representations.

In this section, we assume that the temporal evolution of the reservoir states follows the leaky integrator model [10]

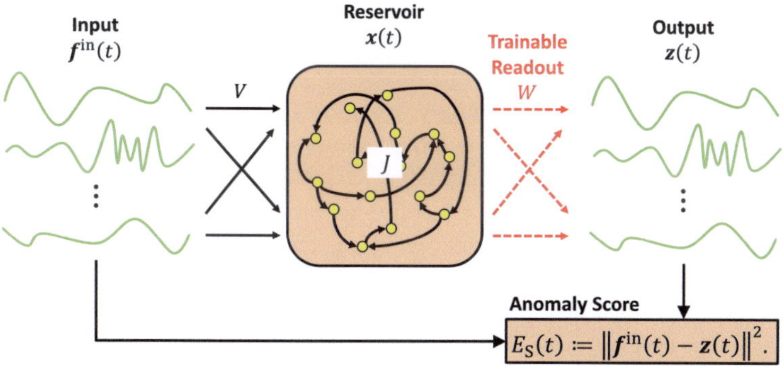

Fig. 1 Reservoir computing approach for time series anomaly detection. Squared reconstruction error $E_S(t)$ is used as an anomaly score

$$\tau \frac{\mathbf{x}(t+1) - \mathbf{x}(t)}{\Delta t} = -\mathbf{x}(t) + J\boldsymbol{\phi}(\mathbf{x}(t)) + V\mathbf{f}^{\text{in}}(t), \tag{1}$$

$$\mathbf{z}(t) = W\boldsymbol{\phi}\left(\mathbf{x}(t)\right). \tag{2}$$

In this formulation, $\mathbf{x}(t) \in \mathbb{R}^N$ represents the reservoir state vector containing activations of N neurons, while τ controls the temporal integration properties. The term $J\boldsymbol{\phi}(\mathbf{x}(t))$ captures recurrent interactions, with $J \in \mathbb{R}^{N\times N}$ defining recurrent connection weights and $\boldsymbol{\phi}(\mathbf{x}) = (\tanh(x_1), \ldots, \tanh(x_N))^\top$ implementing nonlinear activation. External input $\mathbf{f}^{\text{in}}(t) \in \mathbb{R}^L$ influences the reservoir through input weights $V \in \mathbb{R}^{N\times L}$, and the system produces output $\mathbf{z}(t) \in \mathbb{R}^M$ via readout weights $W \in \mathbb{R}^{M\times N}$.

A distinguishing feature of RC is its training efficiency: only the readout weights W are optimized while recurrent (J) and input (V) connections remain fixed. These fixed weights are randomly initialized–J from a Gaussian distribution $\mathcal{N}(0, g^2/N)$ with strength parameter g, and V from uniform distribution $\mathcal{U}(-1, 1)$ with sparsity q. Typically, least squares optimization [8] adjusts W to approximate target outputs $\mathbf{f}^{\text{out}}(t)$ during the training period $t \leq T_0$.

For anomaly detection applications [11–16], RC systems are commonly configured as autoencoders, where the target output equals the input (i.e., $\mathbf{f}^{\text{out}}(t) \equiv \mathbf{f}^{\text{in}}(t)$), as described in Fig. 1. Since the training process primarily exposes the system to normal patterns, the network develops proficiency in reconstructing normal inputs while struggling to accurately reproduce anomalous patterns.

This reconstruction capability forms the basis of anomaly scoring: when presented with normal data, the system generates outputs closely matching the inputs, resulting in small reconstruction errors. Conversely, anomalous inputs produce larger errors due to the network's inability to reconstruct unfamiliar patterns. The squared reconstruction error $E_S := \|\mathbf{f}^{\text{in}}(t) - \mathbf{z}(t)\|^2$ provides a straightforward and effective anomaly score, as illustrated in Fig. 1.

2.3 Application to Handwriting Authentication

Beyond traditional time series anomaly detection scenarios, this section introduces a handwriting authentication task, demonstrating that our slow electronics approach can be applied to human-scale temporal pattern recognitions. This practical example highlights how our method can be applied to biometric security, where the system must recognize legitimate variations in a person's handwriting while rejecting impostor attempts.

In the experiment shown in Figs. 2 and 3, we establish an authentication system where Person A (the authorized user) draws a simple triangle shape on a touchscreen multiple times during the training phase. The system records the x and y coordinates as a bivariate time series, capturing the unique spatiotemporal characteristics of the individual's drawing style, including velocity profiles and natural variations in execution.

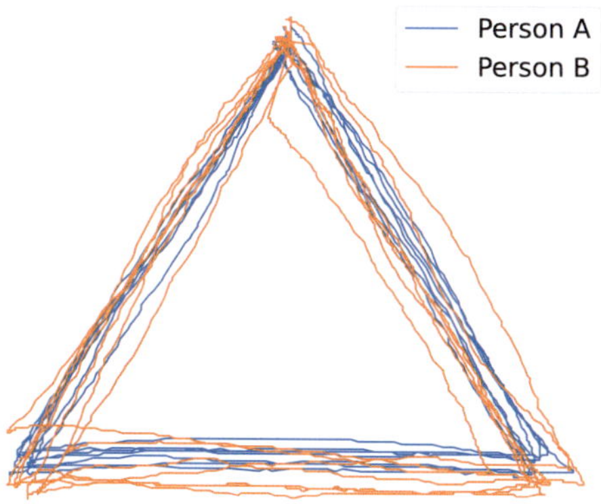

Fig. 2 Handwriting trajectories of two people used in our experiment (test data)

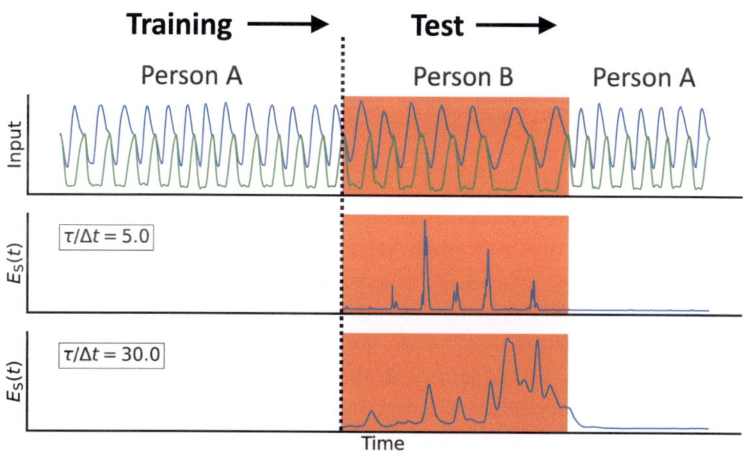

Fig. 3 Handwriting authentication using anomaly detection with reservoir computing. The top panel shows bivariate time series (x,y coordinates) of triangle drawings. Orange highlighting indicates segments that should be detected as anomaly or impostor. The middle and bottom panels show anomaly scores from reservoirs with $\tau/\Delta t = 5.0$ and 30.0 respectively

During the testing phase, both Person A and Person B (an impostor) attempt authentication. The system must correctly identify Person B's attempts (highlighted in orange) as anomalous while accepting Person A's legitimate attempts. This represents a practical biometric verification scenario where the authentication system must distinguish subtle differences in handwriting dynamics that may not be immediately apparent to visual inspection.

The middle and bottom panels in Fig. 3 illustrate the performance of reservoir computing models with different time constants. The parameter $\tau/\Delta t$ represents the ratio of the model's time constant to the sampling interval, effectively controlling how long information persists in the reservoir.

With a smaller time constant ($\tau/\Delta t = 5.0$, the middle panel), the anomaly score $E_S(t)$ shows sharp, localized peaks during Person B's attempts. While these peaks correctly identify moments of significant deviation from Person A's writing patterns, they fail to maintain large scores throughout the entire impostor segment. This results in inconsistent detection that could potentially allow occasional false acceptances during authentication.

In contrast, the model with a larger time constant ($\tau/\Delta t = 30.0$, the bottom panel) produces more sustained large anomaly scores throughout Person B's attempts. This persistent response provides more reliable rejection of the impostor, as the system maintains awareness of the anomalous pattern over longer time periods. The temporal integration properties of the slower reservoir allow it to accumulate evidence of deviation across multiple time points, resulting in more robust detection.

These results demonstrate the importance of appropriate time constants in temporal pattern recognition tasks. The benefits of slower dynamics in the reservoir align with the fundamental concept of slow electronics, where longer time constants enable more effective processing of human-scale temporal patterns. For applications like handwriting authentication, where patterns evolve over seconds rather than milliseconds, reservoirs with slower dynamics can better capture the relevant temporal features.

However, it is worth noting that excessively large time constants can also introduce drawbacks. As seen in the bottom panel, the large anomaly score persists slightly beyond the end of Person B's segment, potentially leading to false rejections when legitimate users attempt authentication immediately after an impostor. This illustrates the need to carefully balance the benefits of temporal integration against the responsiveness of the system.

Finally, even with a simple geometric shape like a triangle, our reservoir computing approach with appropriately tuned time constants can effectively capture the unique temporal dynamics of individual handwriting patterns. For more complex patterns such as signatures or cursive writing, we would expect even greater discriminative performance. This suggests promising applications in biometric security systems, digital signature verification, and other authentication contexts where behavioral biometrics play a crucial role.

3 Emulating a Spiking Neural Network

One of the most well-known nonlinear systems is the biological nervous system, where numerous neurons form neural networks to process information. These neurons are interconnected via synapses and communicate by emitting spikes. The complex interactions between neurons make the neural network a nonlinear and dynamic system, making it an ideal candidate for constructing a reservoir. A neural network that utilizes spikes for signaling is called a spiking neural network (SNN), which is more energy-efficient than the rate-coding implementations described in previous sections because it employs temporally discrete pulse signals.[1]

The behavior of an SNN can be simulated using a computational model. The first step is to emulate the behavior of individual neurons. Here, we focus on the leaky-integrate-and-fire (LIF) model, described in Sect. 3. This model is characterized by the cumulative integration of input spikes (the "integrate" behavior) followed by the gradual relaxation of the membrane potential (V_{M}) back to its resting state (the "leak" behavior). The membrane potential dynamics are governed by the differential equation:

$$\begin{cases} \tau_{\mathrm{n}} \frac{dV_{\mathrm{M}}(t)}{dt} = -V_{\mathrm{M}}(t) + i(t) & ((t - t_{\mathrm{s}}) \geq T_{\mathrm{r}}) \\ V_{\mathrm{M}}(t) = 0 & ((t - t_{\mathrm{s}}) < T_{\mathrm{r}}), \end{cases} \tag{3}$$

where τ_{n} is the neuron's time constant, T_{r} is the refractory period, and $i(t)$ represents the neuron's input. When V_{M} exceeds a critical threshold (V_{C}), the neuron generates a spike, described as

$$s(t) = \begin{cases} 1 & ((t - t_{\mathrm{s}}) < T_{\mathrm{s}}) \\ 0 & ((t - t_{\mathrm{s}}) \geq T_{\mathrm{s}}), \end{cases} \tag{4}$$

where T_{s} is the spike duration, and t_{s} is the time of the last spike. A typical LIF neuron response to pulse inputs is shown in Fig. 4a.

Similarly, the transient response of a synapse upon receiving a spike is approximated by an exponential decay, modeled by the differential equation:

$$\tau_{\mathrm{s}} \frac{dr(t)}{dt} = -r(t) + \frac{s(t)}{T_{\mathrm{s}}}, \tag{5}$$

where $r(t)$ represents the time-dependent post-synaptic current, and τ_{s} is the synaptic time constant. Each output spike $s(t)$ from a neuron is fed into a synapse as its input. The typical transient response of the exponential synapse model is shown in Fig. 4b.

A spiking neural network is constructed using LIF neurons and exponential synapses. Here, we focus on constructing an SNN-based reservoir. The state of the reservoir at time t is described by the so-called response trajectory, denoted as $\mathbf{x}(t) \in \mathbb{R}^N$, which is a vector containing all synaptic responses $r(t)$ in the reservoir, where N is the number of neurons. The outputs from the synapses, represented as $\mathbf{x}(t)$, are mapped to neuron inputs $\mathbf{i}(t)$ by the equation:

$$\mathbf{i}(t) = q(gW\mathbf{x}(t)^{\top} + H\mathbf{i}_{\mathrm{e}}(t)^{\top}) + \mathbf{i}_{\mathrm{b}}, \tag{6}$$

where $\mathbf{i}(t) \in \mathbb{R}^N$ represents the input currents to all neurons in the reservoir, $\mathbf{i}_{\mathrm{e}} \in \mathbb{R}^M$ is the external input, M is the number of input signals, and $\mathbf{i}_{\mathrm{b}} \in \mathbb{R}^N$ is the constant

[1] The total active time of the system is shorter compared to rate-coding implementations, where the system remains continuously active.

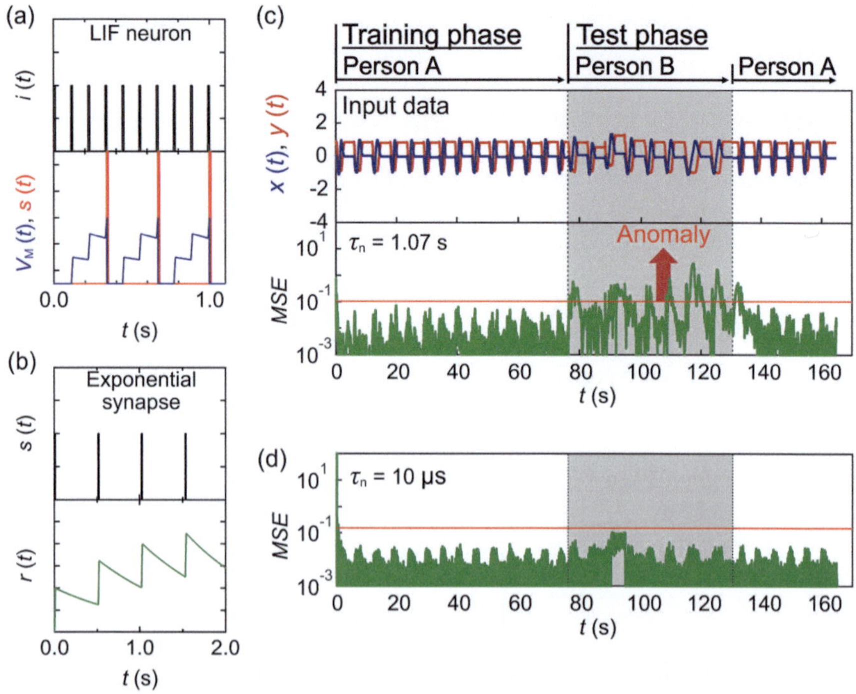

Fig. 4 Emulation of a spiking neural network. **a** Typical output of the LIF neuron model in response to pulse input. **b** Typical output of the exponential synapse model in response to pulse input. **c** Spiking neural network-based handwriting authentication when $\tau_n = 1.07$ s. **d** Same as **c**, but with $\tau_n = 10\,\mu$s. Adopted from [17]

bias input. The matrix $W \in \mathbb{R}^{N \times N}$ describes the connectivity between neurons, while $H \in \mathbb{R}^{N \times M}$ represents the input connection matrix. The parameters q and g define the external and internal signal gains, respectively. The values of q, g, W, and H are hyperparameters that must be optimized for specific tasks.

As a demonstration of SNN-based reservoir computing, we apply the model to handwriting authentication and compare its performance to a rate-coding implementation described in Sect. 2.3. Figure 4c presents the results using a reservoir of 256 neurons and 256 synapses, where the neurons and synapses share the same time constant $\tau_n = \tau_s = 1.07$ s. The hyperparameters used were $W \in \{-2\sigma^2, -\sigma^2, 0, \sigma^2, 2\sigma^2\}^{N \times N}$, where σ^2 is the variance of the values in W, $H \in \{-0.5, 0, 0.5\}^{N \times 2}$, $T_r = 1$ ms, $T_s = \Delta t = 0.1$ ms, $q = 20000$, and $g = 0.001$.[2] The elements of W followed a normal distribution with mean 0 and variance $1.25/N$, and 25 % of the elements in H were nonzero, drawn from a uniform distribution with a mean of 0. The authentication metric defined by

[2] The parameters q and g were adjusted to regulate the firing rate of neurons and ensure $|\mathbf{x}(t)| \lesssim 1$.

$$MSE = (x - \hat{x})^2 + (y - \hat{y})^2, \quad (7)$$

where x and y are the input coordinates, respectively, and $\hat{x}$ and $\hat{y}$ are the output coordinates, respectively, exceeded the threshold value only when the handwriting sample was from a person different from the one used in the training phase, demonstrating successful handwriting authentication.

Interestingly, the accuracy of handwriting anomaly detection depends on the time constants of the neurons and synapses in the reservoir. Figure 4d presents results obtained when $\tau_n = \tau_s = 10$ µs. The hyperparameters used were $T_r = 1$ µs, $T_s = \Delta t = 0.1$ µs, $q = 100$, and $g = 0.125$. In this case, handwriting anomaly detection failed because the neural elements operated at time scales much shorter than those of the input signals. This highlights the importance of slow time scales in human-interactive neural computing.

4 Application to Blood Glucose Prediction

Blood glucose level is an essential parameter for the prevention of cardiac disease, and it rises after eating and decreases with the secretion of insulin [18]. However, in a pathological state, this blood glucose control is not performed properly and abnormal values are taken, causing damage to the body. Therefore, if blood glucose level can be predicted in the future and medication can be administered when the predicted value exceeds the threshold, this will greatly contribute to the prevention of cardiac disease. The ESN used in this study consisted of 60 nodes and was tasked with inputting time-series data of blood glucose levels measured in real time and continuously outputting blood glucose levels at 10–30 min ahead into the future. The input data to the ESN is the time-series data of analog blood glucose level, and they are assumed to be obtained from a general Continuous Glucose Monitor (CGM) device.

In the analog implementation, each node of this ESN requires a short-term memory function according to the algorithm (not long-term), and volatile devices with a time constant of a few minutes is indispensable. This long time constant is needed because the time constant of the short-term memory in ESN should be in the same order as the time scale of the input time series data to be processed[19]. Therefore, as described in detail in Chap. 3, the protonic devices are exploited for this short-term memory. The protonic device in Fig. 13 is used as the device "R_{WO3}" (red color) in Fig. 15c. Because the protonic device can leaky integrate the input spike signal as electrical conductance, it is used as short-term memory. Each node of the ESN stores analog information in protonic device for a short term and performs nonlinear processing by analog CMOS circuits. As a result, the entire network can perform the prediction task according to the ESN algorithm. The input data to the ESN is the time-series data of analog blood glucose level, and they are assumed to be obtained from a general Continuous Glucose Monitor (CGM) device. It should be noted that this paper uses

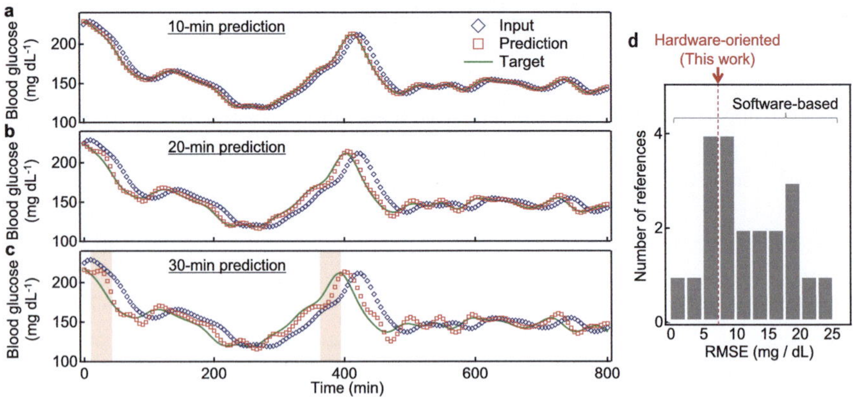

Fig. 5 **a–c** Prediction of 10-min, 20-min, and 30-min ahead blood glucose level, respectively. The prediction value (red square) is well fit by the target value (solid green curve), which is shifted to the left by the prediction time with respect to the input value (blue diamond). The shadowed regions in c clearly shows the prediction can adjust to the sudden onset or fall of the input value. RMSE is 0.62 for 10 min prediction, 3.0 for 20 min, and 7.5 for 30 min [22]. **d** A benchmark of the 30-min prediction of the blood glucose level in comparison with past literature [23]

protonic device for information processing rather than for sensors, although there are some previous studies using nonvolatile resistance-change devices as sensors[20].

The inference results of the blood glucose prediction task utilizing the slow time scale of the protonic device are shown in Fig. 5a-c, where the prediction time was varied from 10 min to 30 min. Here, 10000 min data were separated into 9200 min for training and 800 min for inference. With respect to the input data (blue diamond), the prediction data (red square) is shifted to the left and approximately coincides with the target data (green solid curve), indicating that the prediction tasks were successful. It is important to note that blood glucose levels can change according to diet and exercise by the patient, so it is impossible to make long-term predictions based on the past blood glucose data alone. In other words, Fig. 5b is a short-term prediction with several tens minutes ahead, reflecting blood glucose changes associated with diet and exercise in real time. In fact, in the shaded area in Fig. 5c (30 min prediction), it is seen how the predicted value (red square) are quickly revised to reflect the sudden onset or falling of the input blood glucose levels (blue diamond). While there have been many studies on blood glucose level prediction using various algorithms in general-purpose computers [21], this is the first study to the best of our knowledge to show that the prediction can be achieved even with ultra-low power hardware by using protonic devices as short-term memory and ESN capable of compact training. It should be noted that the prediction performance in this study is comparable to previous studies [21]as shown in Fig. 5d in spite of its hardware-oriented scheme. In Fig. 5d, the Root Mean Squared Errors (RMSEs) is defined for the predicted waveform (ESN output) with respect to the target waveform (30-min shifted input), providing an indicator of prediction quality. The smaller the RMSE is, the better the prediction performance is. Figure 5d plots the number of references as a function of RMSE for 30-min blood glucose prediction. The prediction performance of RMSE = 7.5 (red arrow) is comparable to previous studies while it also satisfies the low power requirement in hardware design.

5 Conclusion

This chapter has demonstrated through numerical simulations that slow electronics can effectively process temporal patterns in practical applications. Our results from handwriting authentication showed that reservoirs with longer time constants ($\tau/\Delta t \approx 30.0$) capture individual writing dynamics better than faster systems, enabling reliable biometric verification. The comparison between different time constants in both rate-coding and spiking implementations confirmed that matching the reservoir's temporal properties to the application's time scale is crucial for optimal performance.

In the blood glucose prediction application, we showed that protonic device-based reservoir computing can forecast glucose levels 10–30 minutes in advance with accuracy comparable to conventional software approaches (RMSE = 7.5 for 30-min prediction), while potentially offering significant power savings. This approach opens promising avenues for ultra-low power computing in domains where longer temporal integration is beneficial, such as healthcare monitoring and biometric security. These simulations establish that slow electronics, when properly matched to application requirements, can effectively process human-scale temporal patterns. The longer time constants of these systems can be advantageous for applications where rapid temporal processing is unnecessary. This approach opens promising avenues for ultra-low power computing in domains where longer temporal integration is beneficial, such as healthcare monitoring and biometric security.

Acknowledgements This work is supported by Japan Science and Technology Agency (JST) CREST Grant Number JPMJCR19K2, JSPS KAKENHI Grant Number JP24K20855 (HT).

References

1. S. Schmidl, P. Wenig, T. Papenbrock, Proc. VLDB Endowment **15**(9), 1779 (2022)
2. F. Rewicki, J. Denzler, J. Niebling, Appl. Sci. **13**(3), 1778 (2023)
3. L. Correia, J.C. Goos, A.V. Kononova, T. Bäck, P. Klein (2022)
4. C. Wang, Z. Zhuang, Q. Qi, J. Wang, X. Wang, H. Sun, J. Liao, Advances in Neural Information Processing Systems, vol. 36 (2023)
5. J. Song, K. Kim, J. Oh, S. Cho, Advances in Neural Information Processing Systems, vol. 36 (2023)
6. C.Y. Lai, F.K. Sun, Z. Gao, J.H. Lang, D. Boning, Advances in Neural Information Processing Systems **36** (2023)

7. D.V. Buonomano, M.M. Merzenich, Science **267**(5200), 1028 (1995)
8. H. Jaeger, GMD Report **148** (2001)
9. W. Maass, T. Natschläger, H. Markram, Neural Comput. **14**(11), 2531 (2002)
10. H. Sompolinsky, A. Crisanti, H.J. Sommers, Phys. Rev. Lett. **61**(3), 259 (1988)
11. O. Obst, X.R. Wang, M. Prokopenko, in *2008 International Conference on Information Processing in Sensor Networks (ipsn 2008)* (IEEE, 2008), pp. 219–229
12. M. Chang, A. Terzis, P. Bonnet, in *International Conference on Distributed Computing in Sensor Systems* (Springer, 2009), pp. 72–86
13. N. Heim, J.E. Avery (2019). arXiv:1909.01709
14. Q. Chen, A. Zhang, T. Huang, Q. He, Y. Song, Neural Comput. Appl. **32**(8), 3685 (2020)
15. T. Lemos, L.F. Campos, A. Melo, N. Clavijo, R. Soares, M. Câmara, T. Feital, T. Anzai, J.C. Pinto, Comput. Chem. Eng. **155**, 107512 (2021)
16. J. Kato, G. Tanaka, R. Nakane, A. Hirose, Nonlinear theory and its applications. IEICE **15**(1), 183 (2024)
17. H. Inoue, H. Tamura, A. Kitoh, X. Chen, Z. Byambadorj, T. Yajima, Y. Hotta, T. Iizuka, G. Tanaka, I.H. Inoue, Adv. Mater. **37**, 2407326 (2025)
18. E. Ackerman, Bull. Math. Biophys. **27**, 21 (1965)
19. G. Tanaka, T. Matsumori, H. Yoshida, K. Aihara, Phys. Rev. Res **4**, L032014 (2022)
20. S. Ginnaram, J. Qiu, S. Maikap, ACS Omega **5**, 7032 (2020)
21. A. Woldaregay, Artif. Intell. Med **98**, 109 (2019)
22. S.P. Pati, Y. Geng, S. Hamasuna, K. Fujiwara, T. Iizuka, H. Inoue, T. Yajima, Commun. Mater. **5**, 177 (2024)
23. P. P, D.S. M, **9**, 1609 (2017)

Learning and Inference in Slow Electronics: FPGA Emulation and Implementation

Yasushi Hotta and Hisashi Inoue

Abstract Slow Electronics has the potential to process time-series data on the same time scale as human activities while achieving ultra-low power consumption. To realize this technology, it is essential to develop dedicated devices such as neuron elements and circuits with long time constants, requiring long-term research and development efforts. In this chapter, as an initial step toward such long-term technological advancements, we aim to explore the potential application fields of Slow Electronics at an early stage. Rather than relying solely on simulations, we conduct hardware-based verification using an FPGA to investigate its practical applicability. To this end, we temporarily set aside power consumption considerations and focus on implementing and analyzing the design and challenges of reservoir computing, a foundational technology for Slow Electronics. Specifically, this study considers the application of Slow Electronics to handwriting authentication technology. By implementing a reservoir on an FPGA to reproduce handwritten input waveforms, we demonstrate its practical feasibility. Furthermore, we identify challenges associated with hardware-based reservoir computing and discuss the necessity of incorporating Slow Electronics elements to address these issues.

1 Background Leading to FPGA Emulation in Slow Electronics

We are currently living in the era of the Internet of Things (IoT), where everything is connected to the network through the internet. With the development of this technology, a vast amount of information from various aspects of our daily lives has been utilized, significantly improving the convenience of our lifestyles. Specifically,

---

Y. Hotta (✉)
Department of Engineering, University of Hyogo, Hyogo, Japan
e-mail: hotta@eng.u-hyogo.ac.jp

H. Inoue
National Institute of Advanced Industrial Science and Technology (AIST), Tsukuba, Japan
e-mail: hisashi.inoue@aist.go.jp

I. H. Inoue (ed.), *Slow Electronics with Reservoir Computing*,
https://doi.org/10.1007/978-981-96-8383-3_6

we now have access to large volumes of data—ranging from videos and images uploaded to servers through the internet to text-based information on social media. However, due to the sheer volume of such data (commonly referred to as "big data"), it has become impossible for humans to process it manually. This has made algorithms such as machine learning and other data processing tools essential. These algorithms are supported by large-scale graphic processing unit (GPU) servers and parallel computing systems, which harness their computational power to efficiently handle vast datasets. Furthermore, these information processing technologies are crucial in services used by many users [1–4].

On the other hand, data related to an individual's activities and vital signs (mainly gathered through wearable devices) can be highly valuable for providing personalized services. Examples include vital signs such as heart rate and blood pressure for health management, [5] as well as location information and movement patterns for monitoring services or local safety alerts [6]. Additionally, the use of biometric information for secure personal authentication is also a promising application [7, 8]. However, because such data often contains personal identifiers, its public use must be restricted in order to protect privacy and personal information.

Fortunately, the amount of information generated by individuals is significantly smaller compared to the vast data transmitted by societal infrastructure. This makes it feasible for personal edge devices to handle this data. As a result, processing this data on personal edge devices is technically achievable, but its management requires customization specific to everyone. These devices must possess characteristics tailored to individual users, distinguishing them from general mass-produced devices. Such devices must be customized and optimized through user learning and usage history. Thus, the key requirements for such devices are as follows:

1. Computational architectures available on the edge device—Sufficient performance must be provided for processing the necessary data.
2. Energy efficiency—Since many edge devices are battery-operated, their energy consumption must be optimized to ensure long-term, real-time operation.
3. Real-world applicability of the applications—The device must support applications that are closely aligned with the user's actual life and activities.

1.1 Architecture Suitable for Implementation on Edge Devices

Recently, reservoir computing (RC) has gained attention as an architecture to implement machine learning frameworks on hardware with limited computational resources, such as edge devices [9]. RC is a derivative model of Recurrent Neural Networks (RNNs), and it could process time-series data. The architecture consists of three parts: input layer, which is the input section with fixed connection weight, a reservoir layer, which is a neural network with fixed connection weights, and a readout layer, which is the output section with variable connection weights. In RC, unlike conventional neural networks, there is no need to update the weights of the entire network using backpropagation. Instead, learning can be achieved by adjusting only the weights of the readout layer, significantly reducing the number of learning parameters. Furthermore, the learning of the readout layer weights can be done using computationally simple methods like least squares, contributing to further reduction in computational costs. This makes it feasible to implement practical applications even on low-power devices [10, 11]. Additionally, the elements that form the reservoir layer do not necessarily have to be neurons, and various phenomena exhibiting nonlinear behavior can be utilized [9]. This opens the possibility of implementing hardware dedicated to the reservoir layer, potentially using physical phenomena in the future.

1.2 Approaches to Energy-Efficient Performance Considering Battery-Powered Devices

From a biological perspective, biological neurons do not operate as fast as digital computers. Neuronal activity relies on the depolarization of sodium (Na^+) and potassium (K^+) ions, and the Na^+ ion channels, which are the source of this activity, require a few milliseconds of inactive time before they can be activated again [12, 13]. Despite such speed constraints, biological neural networks can carry out the necessary information processing for our daily activities without any problems. This implies that the slow operation of these elements does not pose an issue for biological functions in natural environments. Attempting to emulate these inherently slow operations with high-speed digital computers results in substantial delays due to frequency division processes and the associated power overhead, leading to unnecessary energy consumption and reduced power efficiency.

On the other hand, biological neural networks operate with a large number of neurons working asynchronously in parallel, and the system functions based on the superposition of the states of these elements at any given moment. For instance, the human brain cortex is believed to contain up to 100 billion neurons and 100 trillion synapses, [14] which are constantly active, while operating at a metabolic power of about 240 kcal/kg/day [15]. This suggests that the individual components, such as neurons and glial cells, consume extremely low power. Therefore, it is believed that a design based on integrating slow elements that operate at the same timescale and power consumption as biological neurons, similar to how biological neural networks function, would be the ideal approach in neuromorphic devices. This approach is currently being explored in silicon electronics. However, directly implementing the structure of the brain into a silicon chip remains a challenging task [16].

1.3 Hardware Implementation of Required Performance for Real-Time Processing of Time-Series Data

In this chapter, we adopt spiking neurons for the reservoir layer. Spiking neurons are a model that phenomenologically mimics the spike firing of biological neurons, and as such, their behavior is biologically valid. In biological neural networks, neurons process information not with continuous signals but with discrete spikes (firing signals) and relaxing signals via synapses. This feature allows the system to execute operations only when a spike event occurs, without the need to constantly compute the activity of all neurons, unlike conventional neural networks. This significantly reduces memory usage and memory access time in hardware implementations. Although many spiking neuron models have been proposed, [17] we use the Leaky Integrate-and-Fire (LIF) neuron, which is simple yet highly effective. It has been reported that replacing convolutional neural networks with LIF neurons can maintain similar performance in tasks such as MNIST classification [18]. Additionally, using spiking neurons in hardware highly enables energy-efficient implementations by more faithfully replicating the dynamics of the brain's neural network, with very low power overhead [19]. Spiking neural network (SNN) architectures, such as Intel's Loihi, IBM's TrueNorth, and SynSense's DynapCNN, have already been developed.

1.4 Purpose of FPGA-Based Emulation

The contents of this chapter are to develop slow electronics technologies that can perform real-time learning and inference on edge devices for various activities in our daily lives. "Real-time learning" refers to the method of sequentially processing incoming data and simultaneously progressing with learning based on that data. This approach is particularly useful in scenarios where personal information and vital data are continuously generated, and there is a need to process this data on edge devices without sending it to a server.

Furthermore, in interactions between individuals and their environments, there is often no strict time constraint when it comes to the outcome of actions. For example, actions like picking up a coffee cup or thinking and communicating with others occur on a timescale that humans can easily perceive. Digital computers are designed to execute processing at high speeds, often unnecessarily, while human neural activity operates at a much slower pace. Therefore, algorithms that operate on a "slow element" appropriate for the time scale of the target phenomena are crucial in such applications. Systems using slow elements can significantly reduce overhead caused by unnecessary waiting times, offering more efficient processing. Additionally, if the system operates stand-alone, the burden related to network connections and security processing is also reduced.

Table 1 compares the advantages and disadvantages of computer simulation, FPGA, and Application Specific Integrated Circuits (ASIC) implementation. In

Table 1 Comparison of characteristics between computer simulation, FPGA, and ASIC

Aspect	Simulation	FPGA	ASIC
Flexibility for design change	High	Medium	Low
Parallel processing capability	Low	Medium	High
Real-time performance	Low	Medium	High
Development cost	Low	Medium	High
Required expertise	Low	Medium	High

general, computer simulation is highly flexible, allows easy design changes, and has lower development costs. Additionally, development requires minimal specialized knowledge beyond proficiency in programming languages. On the other hand, ASICs are often used for final, mass-production-level implementations where performance and efficiency are critical, but flexibility is not a priority. Furthermore, circuit design for ASICs requires specialized expertise.

FPGA combines characteristics of both programmable logic devices (ex. microcontroller) and ASICs, making it a hybrid of the two. When considering the advantages of using FPGA compared to computer simulation, one of the most significant strengths is its ability to perform parallel processing at the hardware level. FPGAs contain a large number of independently functioning logic cells, which allow tasks to be executed in parallel at the circuit level. Furthermore, FPGA is an large-scale integration (LSI) implemented using existing silicon technology, offering extremely high operational reliability and ensuring data reproducibility. This makes FPGA well-suited for handling parallel tasks with very high speed and real-time performance. The scale of the system is limited by the number of logic cells, so implementations that exceed the available resources cannot be realized. However, FPGA sizes have been steadily increasing, expanding these limits. For example, when one of authors first started using FPGAs in 2006, Xilinx's Virtex-5 series had around tens of thousands of cells, but now, with the Kintex UltraScale series, FPGAs with over a million cells are available. Although FPGA lags behind CPUs and GPUs in some respects, it also has the advantage of benefiting from the performance improvements of mass-produced devices [20].

On the other hand, significant advantage of FPGA compared to ASIC is its ability to easily iterate on experimental designs. Despite being hardware-based, FPGA offers a platform with a balanced trade-off between design flexibility and development cost. This flexibility allows for quick modifications to the hardware design, facilitating rapid prototyping and iterative optimization of algorithms and architectures. By testing directly on hardware, one can make improvements in real-time, advancing the optimization of both the algorithm and system architecture. This flexibility is especially advantageous when validating new architectures or algorithms, making FPGA a powerful tool for experimental design and prototyping.

This chapter aims to investigate the feasibility of developing slow electronics devices by first conducting emulations using FPGA. It is important to clarify that, at this stage, we are not incorporating new material devices or specially developed ASICs into the reservoir or readout layers. These specialized elements are planned to be developed as hardware for slow electronics in the future, at which point some parts of the FPGA emulator will be replaced with these components. Thus, currently, the FPGA-based emulation and digital computer simulation operate on the same basic principles. Moreover, in terms of energy efficiency, it should be noted that overheads from components other than the FPGA—such as processors, I/O controllers, LAN controllers, and working memory—can still be significant. Nevertheless, by emulating the reservoir architecture of SNNs on FPGA, we aim to verify the task-level capabilities that can be executed on existing devices and design a framework for the implementation of future specialized slow electronics hardware. Additionally, by leveraging attractor dynamics on SNNs, we also aim to realize a platform capable of real-time learning and decision-making on-site, without the need for massive logical computations. As a demonstration of this technology, we initially plan to implement handwriting reconstruction system on FPGA.

2 Integrated Development Environment of FPGA

FPGAs are highly flexible and powerful devices, but their development requires proficiency in hardware description languages (HDL), which demands a certain level of experience. However, in recent years, a wide range of development tools and modules has emerged, enabling developers to select the most suitable environment for their specific needs. Examples include high-level synthesis tools that generate HDL code from languages such as C or C++, development environments such as Verilog or VHDL, tools provided by chip vendors (e.g., Vivado by AMD, [21] Quartus by Intel [22]), and software development environments like Vitis IDE by AMD [23]. In this study, we have chosen LabVIEW and the LabVIEW FPGA module as our development platform [24].

LabVIEW is a graphical programming language widely used across various fields, including data acquisition, signal processing, image processing, instrument control, and automated test systems. One of its key features is the ability to manage programs in the form of Virtual Instruments (VIs), which makes it particularly intuitive for FPGA design. In FPGA development, LabVIEW allows users to design circuits as VIs, which can then be automatically synthesized, placed, routed, and converted into programming files (bitstreams). While hardware design typically involves constraints specific to hardware languages that are not present in conventional programming, the LabVIEW FPGA module provides pre-built controls that are tailored to these constraints, enabling developers to adhere to these requirements effortlessly. As a result, one of the most complex aspects of FPGA design—circuit design—becomes more accessible through graphical programming, eliminating the need to write text-based HDL code. However, it is important to note that this module may not be suitable for designing at abstraction levels lower than register transfer level, as shown in Table 2. Specifically, when designing FPGAs from a more fundamental operational level, this tool alone may not be sufficient. Therefore, it is necessary to use this tool in combination with other tools for such designs.

Table 2 FPGA Implementation Levels ranging from very high-level architectural design to very low-level transistor-based representations

HDL description level	Abstraction level	Description
Architecture level	Very high	Level representing system functions such as pipelines and caches
Behavior level	High	Level representing the operation and behavior of circuits
Register transfer level	Middle	Level representing the operation of registers
Gate level	Low	Level representing by logical gate circuit
Switch level	Very low	Level at the component level, such as transistors

Additionally, by using the LabVIEW FPGA module, developers can integrate the development of FPGA device programs with the creation of the user interface (UI) and control systems for the application. This seamless integration allows for close coordination between the logic running on the FPGA and the application controlling or monitoring that logic. Moreover, once a programming file has been generated, it can be downloaded to the FPGA device and executed. The FPGA module can also reprogram the device, allowing for easy modifications and updates even after the device has been implemented. This provides a high level of flexibility in development and allows for continuous adaptation of the FPGA-based system.

However, it is essential to note that the LabVIEW FPGA module is compatible only with FPGA devices mounted on NI's PXI-standard boards. This constraint limits the selection of FPGAs, but the boards in question are equipped with Xilinx's Kintex series FPGAs, which offer excellent performance. For specific examples, refer to Table 3, which shows the reconfigurable boards used in this chapter along with their specifications [25–27]. These boards are equipped with Xilinx's Kintex series FPGAs. Additionally, Table 4 summarizes the comparison between the FPGAs mounted on these boards [28, 29] and that mounted on a typical Zynq board [30]. These FPGAs enable the implementation of the algorithms discussed in this study, and they can serve as a useful benchmark when considering FPGA implementation.

3 Overall Architecture of FPGA RC

This section focuses on the FPGA implementation of RC model with multi-node input and output layers, detailing methods for realizing each component that constitutes these layers. Additionally, it discusses the implementation of parallel processing aimed at approximating the operation of biological neural systems, as well as the potential replacement of components using specialized functional devices and ASICs.

Table 3 Specifications of the NI PXIe-7862, PXIe-7868, and PXIe-7915 FGPA boards

Board model	FPGA	DRAM	Analog input	Sampling rate	Analog Output	Digital Input/ output (DIO)	DIO level
NI PXIe-7862	Kintex-7 325 T	2 GB Single bank	16	1 MS/sec	8	32	3.3 V
NI PXIe-7868	Kintex-7 325 T	2 GB Single bank	8	1 MS/sec	18	32	
NI PXIe-7915	Kintex UltraScale KU060	4 GB Double banks	–	1 MS/sec	–	8	

"-" indicates that the function has not been implemented

3.1 Reservoir Model

Figure 1 illustrates the RC model used as the basis for implementation. This model represents a basic echo state network (ESN) structure comprising an input layer, a reservoir layer, and an output layer, with the input and output layers each containing M nodes. Here, $t(= 0, \Delta t, 2\Delta t, \cdots)$ represents discrete time, and the operations are performed on the hardware at the intervals of computational step Δt. In this chapter, discussion will be developed based on this discrete-time framework. In addition, at $t = 0$, only the reservoir state $\boldsymbol{r}(0)$, determined by the system's initial conditions, is computed, while the input $\boldsymbol{u}(0)$ and output $z(0)$ are not evaluated. These computations begin at $t = \Delta t$, using $\boldsymbol{r}(0)$. The input signal $\boldsymbol{u}(t)$ is adjusted by input nodes and input-layer weights before being projected onto each node of the reservoir layer. The reservoir layer is composed of a network of N mutually connected nodes, employing current-based LIF neuron models as its fundamental elements.

Figure 2 schematically illustrates the transmission of spikes between LIF neurons in the RC. The LIF neurons are interconnected via synaptic filters and network connection weights. When the membrane potential of the preceding neuron exceeds the threshold θ, the neuron fires, and the spike potential is transmitted to the subsequent neuron as a synaptic current through the synapse. In the subsequent neuron, the membrane potential is determined by the postsynaptic current (PSC), which is the sum of all currents received from the preceding neurons, each multiplied by its respective synaptic weight. If the membrane potential of the subsequent neuron exceeds its threshold, it will further propagate signals to the next neuron. In this figure, the black circles represent synaptic weights applied to the input signals, while the white circles indicate synapses. The signals propagate from the synapses to the weights, affecting the subsequent neurons (as shown in this figure, the signal flows from left to right).

Table 4 Specifications of the Kintex-7, Kintex UltraScale, and Kintex UltraScale + series

Part number		Kintex-7 XC7K325T	Kintex ultra scale KU060	Kintex ultra Scale+ ZU9EG
Logic resources	System logic cells	326,080	726,000	599,550
	CLB flip-flops	407,600	663,360	548,160
	CLB LUTs	–	331,680	274,080
Memory resources	Maximum distributed RAM (Kb)	4,000	9,180	8,800
	BRAM/FIFO w/ ECC (36 Kb each)	445	1,080	–
	BRAM/ FIFO (18 Kb each)	–	2,160	–
	BRAM blocks	–	–	912
	Total BRAM (Mb)	16.02	38.0	32.1
Clock	CMT (1MMCM, 2PLLs)	10	12	4
	I/O DLL	–	48	–
I/O Resources	Maximum single-ended I/O	500	–	–
	Maximum differential I/O Pairs	240	–	–
	Maximum single-ended HP I/Os	–	520	–
	Maximum differential HP I/ O pairs	–	240	–
	Maximum single-ended HR I/Os	–	104	–
	Maximum differential HR I/O pairs	–	48	–
Integrated IP resources	DSP slices	840	2,760	2,520
	System monitor	–	1	2
	PCIe®	1 Gen2	3 Gen1/2/3	0

(continued)

Table 4 (continued)

Part number		Kintex-7 XC7K325T	Kintex ultra scale KU060	Kintex ultra Scale+ ZU9EG
	Analog mixed signal (AMS)/ XADC	1	–	1
	Configuration AES/HMAC blocks	1	–	–
	GTX transceivers	8 12.5 Gb/s	–	–
	GTH transceivers	–	32 (16.3 Gb/s)	24 16.3 Gb/s
	GHY transceivers	–		–

"–" indicates that the information is available at https://www.amd.com/content/dam/xilinx/support/documents/data_sheets/ds891-zynq-ultrascale-plus-overview.pdf (p. 6)

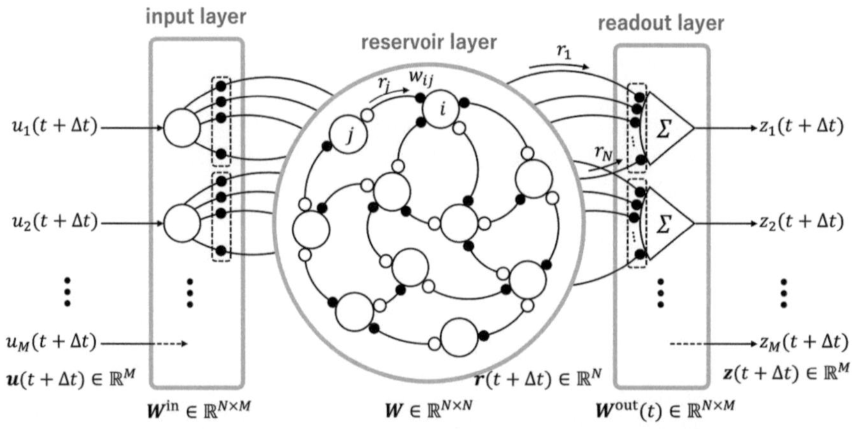

Fig. 1 A typical RC model with multiple inputs and outputs

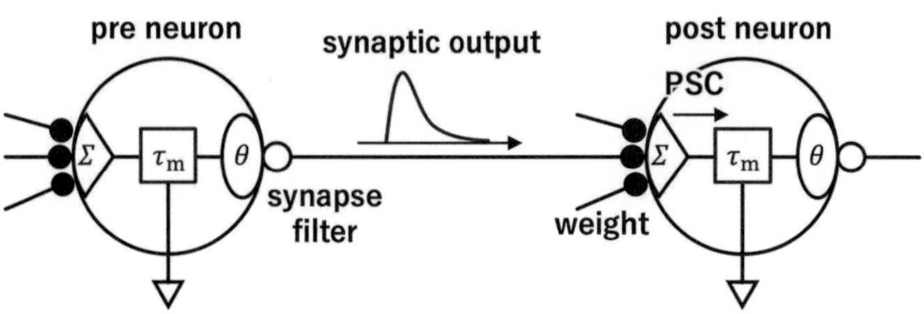

Fig. 2 Two neurons connected via a synapse filter and a synaptic weight in the RC

Referring to Fig. 1, within the reservoir layer, the indices i and j correspond to post-neurons and pre-neurons, respectively. The connection weight from the j-th neuron to the i-th neuron is denoted as w_{ij}, while r_j represents the synaptic current of the j-th neuron. The input current $\boldsymbol{i}_\mathrm{C}$ (t) for each LIF neuron is determined using the following equation:

$$\boldsymbol{i}_\mathrm{c}(t+\Delta t) = \boldsymbol{W}^\mathrm{in} \cdot \boldsymbol{u}(t+\Delta t) + \boldsymbol{W} \cdot \boldsymbol{r}(t) + I_\mathrm{B} \tag{1}$$

Here, $\boldsymbol{W}^\mathrm{in}$ and $\boldsymbol{W}$ represent the input layer weight matrix and the reservoir layer weight matrix, respectively. In RC models, $\boldsymbol{W}^\mathrm{in}$ and $\boldsymbol{W}$ are fixed. $\boldsymbol{W}^\mathrm{in}$ is initialized with elements drawn from a uniform distribution $[-1,1]$ and then scaled appropriately to prevent excessive influence from the input [31, 32]. Additionally, $\boldsymbol{W}$ must be configured to satisfy the Echo State Property. Empirically, it is known that ensuring the spectral radius, i.e., the absolute value of the largest eigenvalue of $\boldsymbol{W}$ is kept below or equal to 1 result in stable system behavior.

The term $\boldsymbol{r}(t)$ is a vector of the synaptic currents filtered through the synaptic filters, while $\boldsymbol{W} \cdot \boldsymbol{r}(t)$ is a vector of the PSCs. Additionally, I_B denotes a bias current that is statically generated within the network.

3.2 LIF Neuron

Physiological neurons fire in response to the frequency of stimuli received from upstream neurons. This firing occurs when the integrated value of the input signals exceeds the membrane potential threshold. On the other hand, if the input stimuli are weak or infrequent, the membrane potential gradually decreases due to leak currents through the cell membrane. As a result, the membrane potential of the neuron naturally decays over time. The LIF neuron model [33] represents the dynamics of membrane potential changes caused by input currents $\boldsymbol{i}_\mathrm{c}(t)$, while simultaneously considering the decay caused by leak currents. In this model, the neuron fires when the increase in membrane potential due to input currents overcomes the decay caused by leaks and exceeds the threshold value. The change in membrane potential $\boldsymbol{v}_\mathrm{m}(t)$ within the LIF neuron, based on the input current received from upstream neurons, is expressed by the following difference equation:

$$\tau_\mathrm{m} \frac{\boldsymbol{v}_\mathrm{m}(t+\Delta t) - \boldsymbol{v}_\mathrm{m}(t)}{\Delta t} = V_\mathrm{rest} - \boldsymbol{v}_\mathrm{m}(t) + R_\mathrm{m}\boldsymbol{i}_\mathrm{c}(t+\Delta t). \tag{2}$$

Here, τ_m is the membrane time constant, V_rest is the resting membrane potential and R_m is the membrane resistance. Rewriting this equation yields the expression for the next step membrane potential as follows:

$$\boldsymbol{v}_\mathrm{m}(t+\Delta t) = \boldsymbol{v}_\mathrm{m}(t) + \frac{V_\mathrm{rest} - \boldsymbol{v}_\mathrm{m}(t) + R_\mathrm{m}\boldsymbol{i}_\mathrm{c}(t+\Delta t)}{\tau_\mathrm{m}} \Delta t \tag{3}$$

This equation demonstrates how the membrane potential can be updated from the membrane potential and input current of the previous time step, as long as the threshold is not exceeded. Using this relationship, the membrane potential at each time step can be sequentially calculated. By determining whether the membrane potential at each time step exceeds the threshold, the firing timing of the neuron can be identified.

$$f_{\theta}(\boldsymbol{v}_{\mathrm{m}}) = \begin{cases} 1, \boldsymbol{v}_{\mathrm{m}} \geq \theta \\ 0, \boldsymbol{v}_{\mathrm{m}} < \theta \end{cases} \tag{4}$$

Here, $f_{\theta}(\boldsymbol{v}_{\mathrm{m}})$ represents the activation function of the threshold θ. In the LIF neuron model, a simple threshold processing method is employed, which uses conditional branching based on the comparison between v_{m} and θ. If v_{m} exceeds θ, a firing spike of the LIF neuron is considered to have occurred, and the spike index is set to 1; otherwise, it is set to 0.

Furthermore, after firing, the neuron's membrane potential is lowered to the reset potential V_{reset} :

$$\boldsymbol{v}_{\mathrm{m}}(t + \Delta t) \leftarrow V_{\mathrm{reset}} \text{ if } \boldsymbol{v}_{\mathrm{m}}(t + \Delta t) \geq \theta \tag{5}$$

After this sequence of processing, a refractory period is introduced to the fired LIF neuron. Specifically, the firing time is set to 0, and the elapsed time step is incremented. During the refractory period, v_{m} remains fixed at the reset potential to prevent leaky integration. Once the refractory period has elapsed, the neuron returns to the leaky integration state. Since this model has a simple structure and most processes are executed through conditional branching, it has low computational cost and is well-suited for hardware implementation.

3.3 Synapse Filter

In physiological neural networks, one of the primary means of signal transmission between neurons is via chemical synapses. A chemical synapse is formed between the axon terminal (terminal button) of the presynaptic neuron and the membrane of the postsynaptic neuron, which may be located on a dendrite, soma, or axon. Between these two neurons exists a minute gap known as the synaptic cleft, which is electrically insulated, preventing the direct flow of electrical current [12].

Signal transmission across a synapse is initiated by the firing (spike potential) of the presynaptic neuron. When the spike potential reaches the axon terminal, voltage-gated calcium ion (Ca^{2+}) channels in the terminal button open, allowing an influx of Ca^{2+} into the cell. The increase in intracellular Ca^{2+} concentration triggers the release of neurotransmitters (e.g., glutamate, GABA) stored in synaptic vesicles. These neurotransmitters diffuse across the synaptic cleft and bind to receptors on the postsynaptic cell. This receptor binding induces the opening and closing of ion channels in the postsynaptic membrane, leading to the influx or efflux of ions such as Na^+, K^+, and chloride (Cl^-). Consequently, the membrane potential of the postsynaptic neuron changes, generating either an excitatory or inhibitory PSC [12].

The temporal dynamics of the postsynaptic current exhibit characteristic behavior, where the current rapidly increases following spike generation and then gradually decays. A mathematical model that captures this behavior is the double-exponential synapse. Rather than rigorously reproducing the detailed physiological mechanisms of actual synapses, this model focuses on efficiently approximating the temporal dynamics of the postsynaptic current. Due to its simplicity and ease of implementation, the double-exponential synapse is widely employed in SNNs [34].

The $\boldsymbol{r}(t)$ of the double-exponential synapse can be obtained using the following two difference equations:

$$\boldsymbol{r}(t+\Delta t) = \left\{1 - \frac{\Delta t}{\tau_d}\right\}\boldsymbol{r}(t) + \boldsymbol{h}_r(t)\Delta t \tag{6}$$

$$\boldsymbol{h}_r(t+\Delta t) = \left\{1 - \frac{\Delta t}{\tau_r}\right\}\boldsymbol{h}_r(t) + \frac{1}{\tau_r \tau_d}\boldsymbol{s}(t) \tag{7}$$

where τ_r is the rise time constant, determining the timescale over which the postsynaptic current rapidly increases, and τ_d is the decay time constant, governing the rate at which the current decays over time. The function $\boldsymbol{s}(t)$ represents the spike activation function of each neuron and is given by:

$$\boldsymbol{s}(t) = f_\theta\{\boldsymbol{v}_{\mathrm{m}}(t)\} \tag{8}$$

which returns 1 when a spike occurs and 0 otherwise. By employing this double-exponential model, the characteristic rapid increase and gradual decay of postsynaptic currents can be succinctly represented. This model is particularly well-suited for FPGA implementations.

3.4 Updating W^{out} Using Recursive Least Squares Method

The output layer generates the output signal $z(t+\Delta t)$ by receiving input from the nodes in the reservoir layer via the output weights $\boldsymbol{W}^{\mathrm{out}}(t)$, as represented by the following equation:

$$z(t+\Delta t) = \boldsymbol{W}^{\mathrm{out}}(t) \cdot \boldsymbol{r}(t+\Delta t) \tag{9}$$

Each time new data is sequentially input and the reservoir state is updated, the output weight matrix $\boldsymbol{W}^{\mathrm{out}}$ is updated accordingly. This update is performed using the Recursive Least Squares (RLS) algorithm [35]. RLS extends the least squares

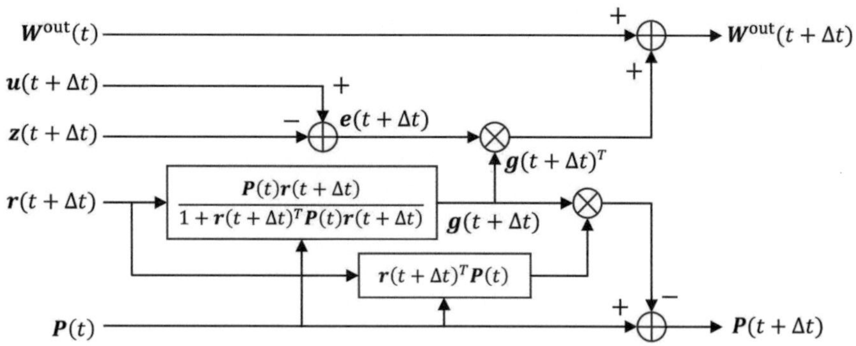

Fig. 3 Block diagram of the processing flow for updating the matrices $\boldsymbol{W}^{out}$ and $\boldsymbol{P}$ using the RLS algorithm

method for parameter estimation, providing an efficient way to update system parameters when data is received sequentially. In systems where past data influences current data, parameter updates must account for prior responses whenever new inputs arrive. However, RLS leverages values from the previous step to compute the updated parameters, significantly reducing computational costs. This makes RLS particularly suitable for real-time applications in dynamic environments.

The following describes the FPGA implementation of the parameter update process using RLS. Figure 3 illustrates the block diagram of the processing flow within a single step of the RLS algorithm. The inputs from the left side—$u(t+\Delta t)$, $r(t+\Delta t)$, and $z(t+\Delta t)$—correspond to signals from the input layer (if the training signal is the input), the reservoir layer, and the output layer, respectively. Additionally, $\boldsymbol{W}^{out}(t)$ represents the output layer weight matrix from the previous step, while $\boldsymbol{P}(t)$ is the inverse of the reservoir state autocorrelation matrix. The vector $\boldsymbol{g}(t+\Delta t)$ is known as the gain vector.

At $t=0$, initial values are set as follows:

$$\boldsymbol{P}(0) = \delta^{-1}\boldsymbol{I} \tag{10}$$

$$\boldsymbol{W}^{out}(0) = \boldsymbol{O} \tag{11}$$

where $\boldsymbol{I}$ is the identity matrix, δ is a constant (typically ≤ 1), and $\boldsymbol{O}$ is the zero matrix. These initial values define the system's starting state, after which the output $z(\Delta t)$ for the first input $u(\Delta t)$ is computed. For each subsequent step after Δt, the update equations are as follows:

$$\boldsymbol{e}(t+\Delta t) = \boldsymbol{u}(t+\Delta t) - \boldsymbol{z}(t+\Delta t) = \boldsymbol{u}(t+\Delta t) - \boldsymbol{W}^{out}(t)\boldsymbol{r}(t+\Delta t) \tag{12}$$

$$\boldsymbol{g}(t+\Delta t) = \frac{\boldsymbol{P}(t)\boldsymbol{r}(t+\Delta t)}{1+\boldsymbol{r}(t+\Delta t)^T\boldsymbol{P}(t)\boldsymbol{r}(t+\Delta t)} \tag{13}$$

$$\boldsymbol{W}^{\text{out}}(t+\Delta t) = \boldsymbol{W}^{\text{out}}(t) + \boldsymbol{e}(t+\Delta t)\boldsymbol{g}(t+\Delta t)^T \tag{14}$$

$$\boldsymbol{P}(t+\Delta t) = \boldsymbol{P}(t) - \frac{\boldsymbol{P}(t)\boldsymbol{r}(t+\Delta t)\boldsymbol{r}(t+\Delta t)^T\boldsymbol{P}(t)}{1+\boldsymbol{r}(t+\Delta t)^T\boldsymbol{P}(t)\boldsymbol{r}(t+\Delta t)} \tag{15}$$

Here, $\boldsymbol{e}(t+\Delta t)$ represents the output error, which is the difference between the input $\boldsymbol{u}(t+\Delta t)$ and the predicted output obtained using the updated reservoir state $\boldsymbol{r}(t+\Delta t)$ and the previous weight matrix $\boldsymbol{W}^{\text{out}}(t)$. The gain vector $\boldsymbol{g}(t+\Delta t)$ is computed based on $\boldsymbol{r}(t+\Delta t)$ and $\boldsymbol{P}(t)$. These values are used to update $\boldsymbol{W}^{\text{out}}(t+\Delta t)$ and $\boldsymbol{P}(t+\Delta t)$, which are then used in the next calculation step.

In the FPGA implementation, the weight update mechanism for the output layer is designed to be modular, allowing it to be decoupled from the reservoir update process via internal state switches. This means that when learning is unnecessary, the update mechanism can be disabled. In practice, rather than updating at every Δt, updates are performed periodically—every few dozen reservoir updates—to balance computational efficiency and system performance.

4 Implementation of RC for SNNs on FPGA

The discussion now shifts to the individual components constituting each layer, and methods for mapping these components to hardware models are examined. In FPGA implementations, it is crucial to define the specific configurations and design objectives for each component at the initial design stage. Unlike software programs executed on PCs, FPGA configurations such as memory capacity and other hardware parameters cannot be dynamically modified after deployment. Although reconfiguration is possible, it requires redesigning circuits and regenerating programming files. Therefore, it is essential to anticipate concrete operating conditions during the initial design phase and allocate resources accordingly. Moreover, this design approach is vital for achieving efficient circuit implementation while conserving resources. While it is technically feasible to allocate large memory capacity during the design stage, this can lead to excessive consumption of slice resources or increased utilization of block random access memory (BRAM), resulting in reduced efficiency in circuit implementation. Ultimately, this may impose limitations on the scale of systems that can be implemented. Consequently, FPGA design requires the establishment of clear objectives in advance, along with appropriate resource allocation, to ensure efficient and practical system implementation.

As a concrete example, this section describes the FPGA implementation methodology of an RC model designed to process handwritten trajectory data obtained from a pen tablet, as shown in Fig. 4. In this system, the trajectory of the pen, captured via tablet input, is treated as planar coordinate data, and serves as the target for learning. Consequently, the model is structured with two input and two output nodes. Furthermore, the number of neurons, denoted as N, is assumed to be at most 2^{10} and is

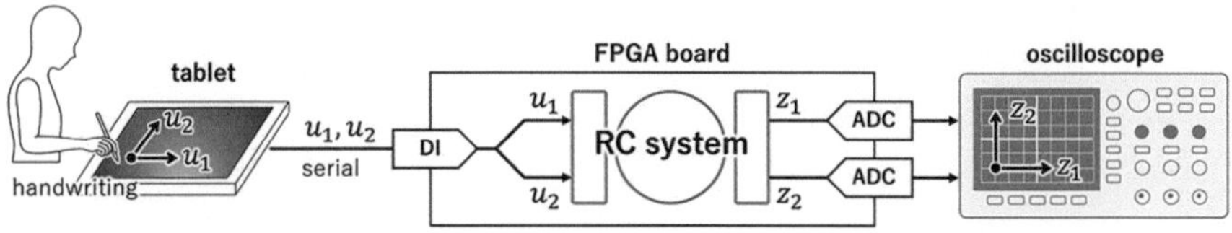

Fig. 4 FPGA-based RC system learning handwriting trajectory data input from a tablet

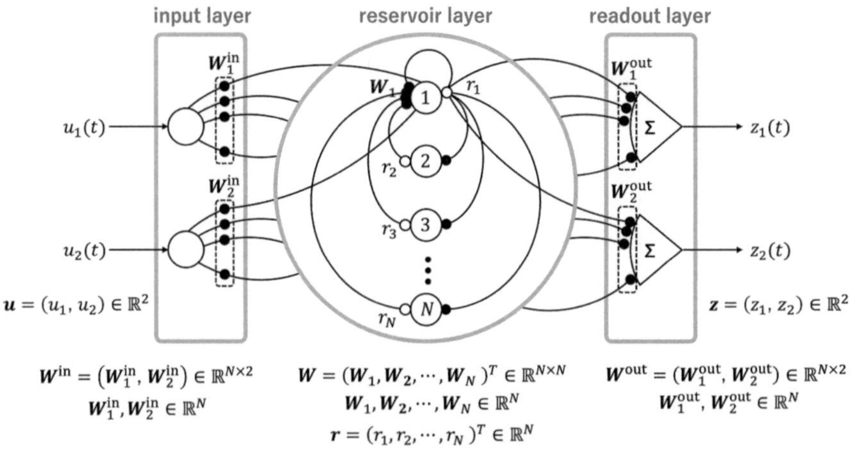

Fig. 5 The specified RC model for processing handwriting trajectory data

designed to be configurable within the range of $N = 1 \sim 2^{10}$. To accommodate this variability, memory allocation must be planned at the implementation stage to support the maximum neuron count of $N = 2^{10}$. Based on these design conditions, an overview of the RC model tailored for learning handwritten trajectory data is illustrated in Fig. 5.

4.1 Implementation of the Reservoir Layer in FPGA

Figure 5 illustrates the reservoir layer with a focus on the input and output of the first LIF neuron to enhance conceptual clarity regarding the implementation. The input to this neuron consists of two components: the weighted sum of inputs from the input layer, multiplied by the input weight matrix $\boldsymbol{W}^{\text{in}}$, and the weighted sum of inputs from other neurons within the reservoir, multiplied by the recurrent weight matrix $\boldsymbol{W}$. The output of this neuron, denoted as r_1, is then propagated to all other neurons in the reservoir as well as to the output layer. Although omitted in Fig. 1, the same computation applies to all N neurons in the system. To systematically describe this system, Eq. (1) can be rewritten in terms of vectors and matrices as follows:

$$\begin{bmatrix} i_{c1} \\ i_{c2} \\ \vdots \\ i_{cN} \end{bmatrix}_{t+\Delta t} = \begin{bmatrix} w^{\text{in}}_{11} & w^{\text{in}}_{12} \\ w^{\text{in}}_{22} & w^{\text{in}}_{22} \\ \vdots & \vdots \\ w^{\text{in}}_{N2} & w^{\text{in}}_{N2} \end{bmatrix} \begin{bmatrix} u_1 \\ u_2 \end{bmatrix}_{t+\Delta t} + \begin{bmatrix} w_{11} & w_{12} & \cdots & w_{1N} \\ w_{21} & w_{22} & \cdots & w_{2N} \\ \vdots & \vdots & & \vdots \\ w_{N1} & w_{N2} & \cdots & w_{NN} \end{bmatrix} \begin{bmatrix} r_1 \\ r_2 \\ \vdots \\ r_N \end{bmatrix}_t + I_B \quad (16)$$

Here, u_1 and u_2 correspond to the dynamic x-coordinate and y-coordinate input data from the pen tablet at $t + \Delta t$, respectively. The subscript after each square bracket represents the discrete-time step. Next, z is described by Eq. (8) in terms of individual elements as follows:

$$\begin{bmatrix} z_1 \\ z_2 \end{bmatrix}_{t+\Delta t} = \begin{bmatrix} w^{\text{out}}_{11} & \dots & w^{\text{out}}_{1N} \\ w^{\text{out}}_{22} & \dots & w^{\text{out}}_{2N} \end{bmatrix}_t \begin{bmatrix} r_1 \\ r_2 \\ \vdots \\ r_N \end{bmatrix}_{t+\Delta t} \quad (17)$$

Through this equation, the system derives outputs z_1 and z_2 corresponding to the inputs u_1 and u_2.

4.2 Handling Weight Matrices in FPGA

In this system, the weight matrices for the input layer, reservoir layer, and output layer are represented as multi-dimensional matrices. However, the FPGA module used in this implementation does not support direct multi-dimensional matrix definitions. Therefore, these matrices must be treated as one-dimensional arrays.

To accommodate this constraint, each neuron's weights are designated as $\boldsymbol{W}_1, \boldsymbol{W}_2, \cdots, \boldsymbol{W}_N$, and these are stored in a one-dimensional structure, as depicted in Fig. 6. In this format, for an element $\boldsymbol{W}_{kl}$ located at row $k(\text{where} k = 0 \,\text{to}\, N-1)$ and column $l(\text{where} l = 0 \,\text{to}\, N-1)$, the corresponding index in a one-dimensional array is given by:

$$\text{Index} = mN + n \quad (18)$$

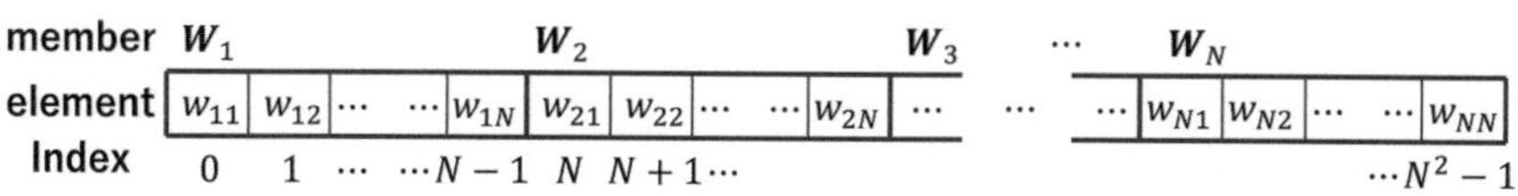

Fig. 6 Schematic diagram of accessing elements in a 2D array using a 1D array with the Index variable

Table 5 Number of elements for each variable when $N = 1024$

Variable	Number of elements	Variable	Number of elements
$\boldsymbol{u}$	2	$\boldsymbol{W}^{\mathrm{in}}$	$2N = 2048$
$\boldsymbol{z}$	2	$\boldsymbol{W}$	$N^2 = 1048576$
$\boldsymbol{e}$	2	$\boldsymbol{W}^{\mathrm{out}}$	$2N = 2048$
$\boldsymbol{i}_{\mathrm{PSC}}$	1024	$\boldsymbol{s}$	1024
$\boldsymbol{v}_{\mathrm{m}}$	1024	$\boldsymbol{g}$	1024
$\boldsymbol{r}, \boldsymbol{h}_r$	1024	$\boldsymbol{P}$	$N^2 = 1048576$

This index mapping allows efficient memory access, facilitating matrix operations such as transposition.

4.3 Determining the Required Number of Elements

The accuracy of the reservoir is known to improve as the number of implemented units increases. In computational systems, constraints on the number of units typically stem from processing time and memory resources. In FPGA implementations, however, the primary limiting factor is often the available memory capacity rather than computational speed.

Suppose we set the number of LIF neurons in the reservoir to $N = 2^{10} = 1024$. In this case, it is necessary to allocate sufficient memory to accommodate all neurons' computations. Furthermore, by pre-allocating memory for all elements, the system allows for dynamic configurability of N within the range of 1 to 1024, even after implementation. Table 1 summarizes the required number of elements for each variable. This allocation ensures that all necessary computations are accommodated while allowing the system to adjust the number of neurons flexibly (Table 5).

4.4 Processing Flow and Memory Access

Figure 7 illustrates the processing flow of RC and the memory access involved in each block. In FPGA implementation, the logical circuits are designed to sequentially execute the processing blocks shown in the figure. Various system states and variables are primarily stored in on-chip memory (BRAM). Meanwhile, large matrix data, such as $\boldsymbol{W}$ and $\boldsymbol{P}$, are stored in on-board memory (dynamic random-access memory: DRAM). In the figure, downward arrows indicate memory write operations, while upward arrows indicate memory read operations.

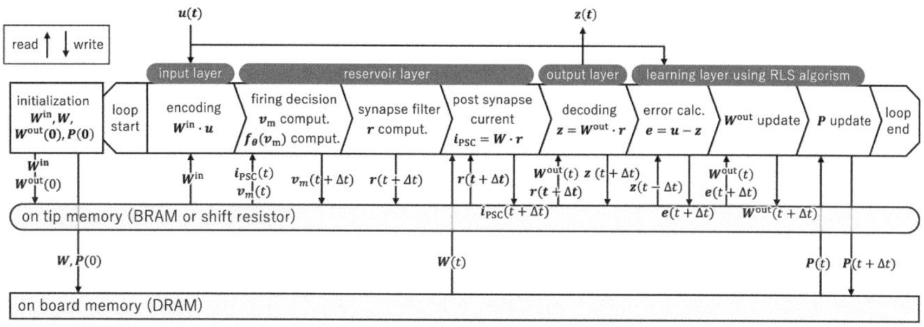

Fig. 7 Flow diagram and illustrating the sequential processing and memory accesses through the input layer, reservoir layer, output layer, and learning layer for the RC on FPGA

The initialization block sets up various parameters, including the input layer weights, reservoir layer weights, network sparsity, and initial values for each parameter. These values are then uploaded to memory. Once initialization is complete, the process enters a loop, where the following computations are repeatedly executed:

1. Input layer computation: Encoding external input data.
2. Reservoir layer computation: Updating the neuron states.
3. Output layer computation: Computing results based on the reservoir layer output.
4. Error calculation: Evaluating the difference between the computed output and the target signal.
5. Weight update: Adjusting $\boldsymbol{W}^{\mathrm{out}}$ and $\boldsymbol{P}$ based on the error.

Since each processing block must use the updated values from the backward block, the computations need to be executed sequentially. However, within each block, pipeline processing can be leveraged to enable parallel execution, thereby reducing overall processing time.

This section presents the results of an analysis of the computation time required for each processing block. The measurement was conducted using a RC system with 1024 LIF neurons and a single input, performing a task to reproduce a sinusoidal waveform. This processing task is similar to the handwritten recognition in this chapter and involves a comparable computational load. If we define one processing unit per FPGA clock cycle as "1 Tick", the total time required for the loop processing in this task was 93.575 M Ticks. Given that the base clock of the FPGA module NI PXIe-7915 is 80 MHz, this corresponds to an actual processing time of approximately 1.1 s (which is considered slow even from the perspective of slow electronics!). The breakdown of computation time is as follows:

- Neuron state update in the reservoir layer: 3.549 M Ticks.
- Output layer and error calculation: 0.026 M Ticks.
- Weight update in the readout layer (RLS algorithm for updating $\boldsymbol{W}^{\mathrm{out}}$ and $\boldsymbol{P}$): approximately 90 M Ticks.

Figure 8 presents a pie chart illustrating the proportion of each process. As evident from the graph, more than 96% of the total FPGA computation time is consumed by weight updates in the readout layer. This highlights the fact that, when implementing online learning, the computational load of the readout layer is particularly significant.

The primary reason for the long processing time in the RLS update procedure is the multiple read and write operations on the matrix $\boldsymbol{P}$ from DRAM within this processing block. Since the size of the matrix $\boldsymbol{P}$ increases proportionally to the square of the neuron count N, the overhead associated with DRAM access becomes significant. The details of this processing block are illustrated in Fig. 9. First, Fig. 9a represents the computation block for the gain vector $\boldsymbol{g}(t + \Delta t)$. In this block, the matrix $\boldsymbol{P}$ is first read from DRAM, after which the product with the reservoir state vector $\boldsymbol{r}(t + \Delta t)$ is computed to obtain the gain vector. Next, Fig. 9b depicts the update process for $\boldsymbol{W}^{\mathrm{out}}(t)$. In this step, $\boldsymbol{W}^{\mathrm{out}}(t)$ is updated using the gain vector and the error vector $\boldsymbol{e}(t + \Delta t)$. However, since this computation does not require direct reference to data stored in DRAM, no DRAM access occurs. Finally, Fig. 9c shows the update of the matrix $\boldsymbol{P}$. This process begins with reading the matrix $\boldsymbol{P}$ from DRAM, followed by the necessary computations based on the update equation. The updated $\boldsymbol{P}$ matrix is then written back to DRAM. Since all elements of the matrix $\boldsymbol{P}$ is accessed during this step, DRAM bandwidth is likely to become a bottleneck. Therefore, the most computationally expensive operations in this update process are those related to the matrix $\boldsymbol{P}$, with the DRAM accesses occurring in Figs. 9a and c being the dominant factors contributing to the overall processing time.

The matrix $\boldsymbol{P}$ contains N^2 elements, making it significantly larger compared to other variables. This characteristic also applies to the reservoir weight matrix $\boldsymbol{W}$. Since such large memory-intensive variables cannot be accommodated within the BRAM of the FPGA, they must be stored in DRAM. However, DRAM access speed is considerably lower than that of BRAM. Moreover, the large number of elements further exacerbates processing time. Here, to evaluate the difference in memory access speeds between DRAM and BRAM, we implemented an FPGA-based module

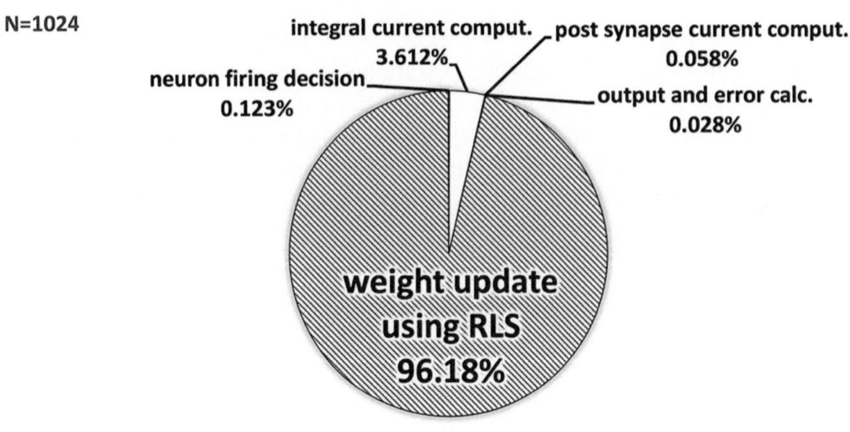

Fig. 8 Pie chart showing the proportion of processing time for each layer during a single computation step Δt (when $\boldsymbol{W}^{\mathrm{out}}$ and $\boldsymbol{P}$ are updated)

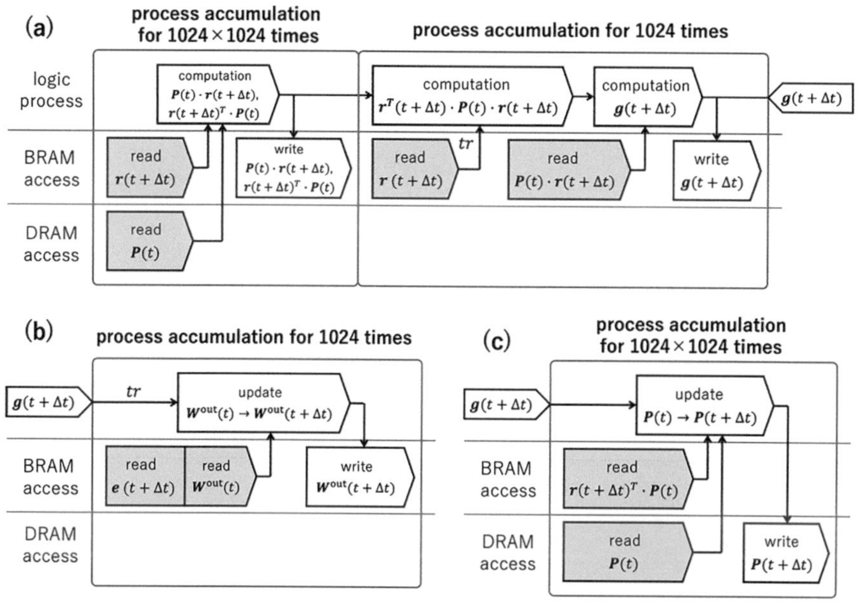

Fig. 9 Summary of the computation flow and accumulated processing times for each step during the update of the matrices $\boldsymbol{W}^{\mathrm{out}}$ and $\boldsymbol{P}$. **a** computation step to obtain the gain vector $\boldsymbol{g}$, **b** updated $\boldsymbol{W}^{\mathrm{out}}$, and **c** updated $\boldsymbol{P}$

that exclusively handles single-precision floating-point memory access operations and measured the number of Ticks required as well as the total processing time. The results are presented in Fig. 10. In this figure, the horizontal axis represents the number of elements, which is proportional to the square of the neuron count N. The left vertical axis indicates the number of Ticks required for memory access (in mega-Ticks), while the right vertical axis corresponds to the processing time (in seconds). As observed in Fig. 10, DRAM access time is significantly longer compared to BRAM, and the Ticks required for DRAM read/write operations increase sharply as the number of elements grows. This finding suggests that DRAM access represents a major computational bottleneck. Conversely, BRAM exhibits significantly shorter access times, maintaining relatively low Tick counts even as the number of elements increases. Nevertheless, it is important to note that the results presented here do not assume an optimized memory access architecture. Techniques such as memory buffering and multi-bank parallel access can further reduce processing time. Even so, the data provided in this study serve as a baseline for understanding the relationship between the number of elements and memory access latency. Furthermore, if the reservoir size does not need to be excessively large, prioritizing the use of on-chip memory (BRAM) can significantly reduce memory access time and improve overall processing efficiency.

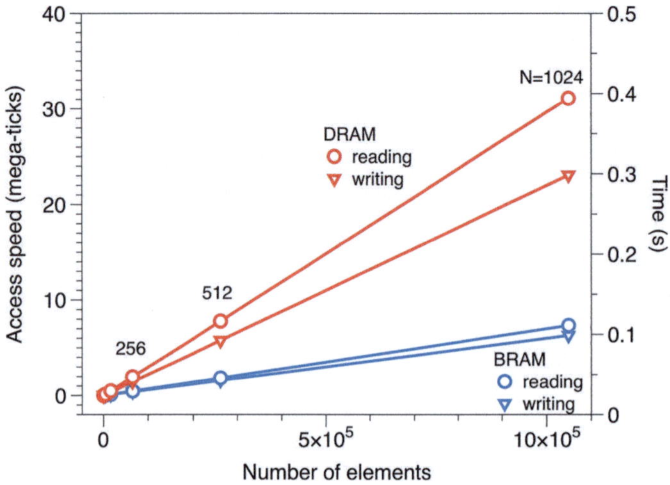

Fig. 10 Comparison of memory access speeds between DRAM and BRAM as a function of the total number of elements (equal to the square of the neuron count N)

4.5 Demonstration of Handwriting Reconstruction System on FPGA

Finally, we present the results of the demonstration of our implemented system for reproducing handwritten input.

Figure 11 shows the external appearance of the tablet used for acquiring handwritten data (a) and the front panel of the system's user interface created with LabVIEW (b). This system learns the input waveform of handwritten strokes on a tablet and generates a reproduced waveform as output. The input data consists of the time-series waveform of pen-tip coordinates obtained from the tablet (Fig. 12b, center), while the output data represents the reproduced waveform generated after learning (Fig. 12b, left). In this demonstration, a triangular shape is drawn four consecutive times as input, with each triangle taking approximately five seconds to complete. Figure 12b (right) presents the trajectory of both the input and output waveforms, synchronized in time. A comparison between the input trajectory (bottom) and the output trajectory (top) confirms that the reproduced output successfully reconstructs the shape of the input triangular strokes.

In the field of slow electronics, applications in personalized services implemented on edge devices are expected. Based on this demonstration, we are currently developing a handwriting-based personal authentication technology. Compared to conventional password authentication, possession-based authentication, and biometric authentication, handwriting authentication is expected to offer advantages in terms of convenience and protection against key theft. Furthermore, we aim to explore the practical feasibility of slow electronics in conjunction with the development of dedicated hardware that enables parallel processing of neurons [36, 37].

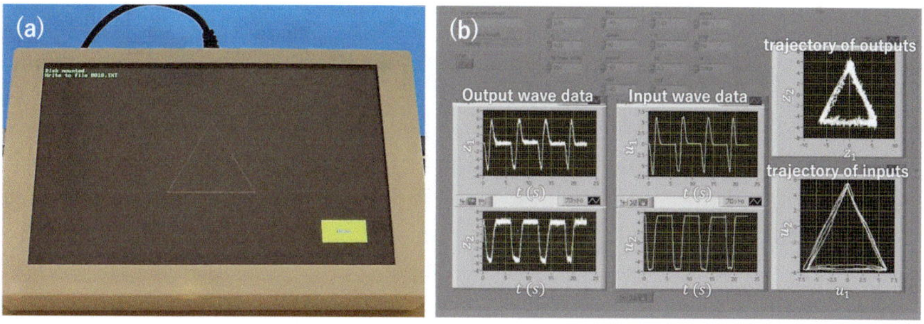

Fig. 11 **a** Handwriting input tablet. The shape displayed on the screen (a triangle in this case) serves as a guideline during input. **b** Front panel image of the user interface for the handwriting reconstruction system. The highlighted areas on the front panel indicate graphs of the dynamic coordinate data for the input (u_1, u_2) and output (z_1, z_2). The darker shaded area contains the system configuration settings

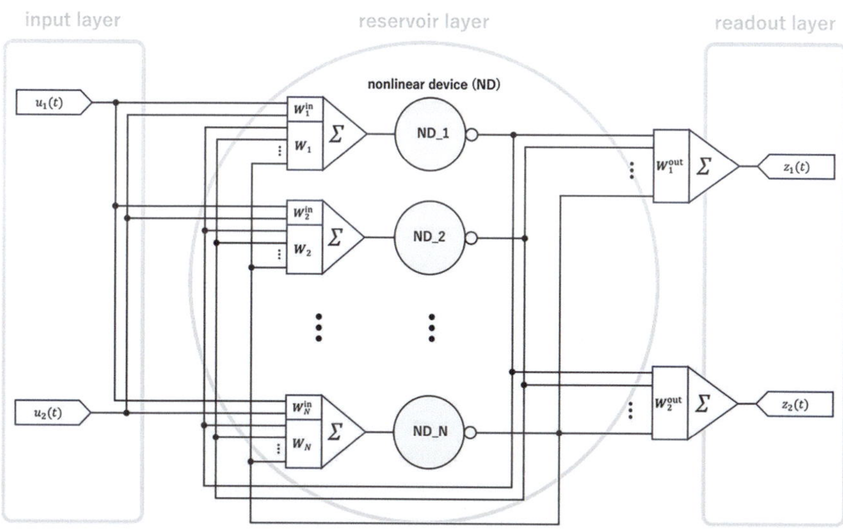

Fig. 12 Architecture model of a RC that reduces memory access dependency by parallelizing nonlinear devices equipped with registers

5 Conclusion and Future Perspectives

We conclude this chapter by proposing a design approach aimed at replacing conventional architectures with dedicated slow electronics devices. In hardware implementations of RC, the most fundamental approach is to implement the necessary number of neuron and synapse elements that operate in a massively parallel manner, similar to the brain, and interconnect them in a network structure. Numerous studies have been conducted to integrate neuron and synapse elements into LSI and reproduce the functional aspects of neural information processing. Notably, efforts using Si LSI are often referred to as Silicon Neuronal Networks (SNN)—an acronym that coincides with Spiking Neural Networks (SNN) [16].

On the other hand, some argue that parallelization is not necessarily an essential solution in the hardware implementation of neural networks. This perspective is based on the idea that with a high-speed processor and large-capacity memory, a high-performance neural network can be implemented even with a general-purpose processing unit, reducing the need for parallelization. Furthermore, the rapid advancement of mass-production devices enables the use of increasingly powerful processors and memory, further supporting this argument [20].

However, from the perspective of slow electronics, parallelization is a fundamental issue. One of the key reasons is that the goal of slow electronics is to limit the operating speed of individual components, thereby reducing unnecessary overhead that arises in high-speed computing systems. By executing only the necessary computations at the necessary timing and in the necessary amount, overhead can be further minimized. Instead of relying on a single high-speed processor, leveraging a large number of low-speed components operating in parallel allows for improved power efficiency while achieving high computational performance. Therefore, in hardware design, it is crucial to carefully consider parallelization strategies and construct an optimal architecture.

In FPGA implementations, if logic circuits could be used without constraints, it would theoretically be possible to achieve full parallelism. To explore this, we attempted to implement Izhikevich model neurons, [38, 39] which closely mimic biological neurons, in parallel on an NI PXIe-7862 (or PXIe-7868) board. However, the maximum number of neurons that could operate in parallel and asynchronously was 40, even without incorporating additional functionalities such as synaptic filters, connection weights, or network coupling. Considering these additional components, resource limitations would become even more severe, making a fully parallel implementation challenging. Therefore, developing physical neuron elements to realize such a parallel architecture becomes a crucial research direction.

In recent years, research on Physical Reservoir Computing (PRC) has been actively pursued, primarily focusing on replacing the reservoir layer [40]. Given the current scale of implementation, where the number of physical elements is relatively small, the processing overhead beyond the output layer is not yet a significant concern. However, as shown in Figs. 8 and 10, as the number of nodes increases, the computational cost of the weight update layer is expected to become a critical bottleneck.

Figure 12 illustrates a conceptual diagram of a parallel architecture. In this design, no dedicated memory is incorporated. Instead, the reservoir states are stored in registers corresponding to nonlinear elements at each node. Moreover, the values obtained by multiplying these states with connection weights are also stored directly in registers, ensuring that all nodes can access their respective data simultaneously and operate in a highly parallel manner. This system configuration minimizes memory accesses, effectively mitigating the memory access bottleneck observed in conventional architectures.

Within this context, the concept of slow electronics is expected to emerge as a key technology driving this field forward. In contrast to conventional approaches that focus solely on achieving higher computational speed, technologies that enable brain-inspired computation with low power consumption will play a fundamental role in establishing the next-generation edge AI hardware that operates efficiently in real-world environments.

References

1. X.-W. Chen, X. Lin, IEEE Access **2**, 514 (2014)
2. A.L. Beam, I.S. Kohane, JAMA **319**, 1317 (2018)
3. S. Dash, S.K. Shakyawar, M. Sharma, S. Kaushik, J. Big Data **6**, 54 (2019)
4. C. Janiesch, P. Zschech, K. Heinrich, Electron. Mark. **31**, 685–695 (2021)
5. J. Chen, X. Ran, Proc. IEEE **107**, 1655–1674 (2019)
6. Z. Wang, Z. Yang, T. Dong, Sensors **17**, 341 (2017)
7. Z. Rui, Z. Yan, IEEE Access **7**, 5994 (2019)
8. K. Bibi, S. Naz, A. Rehman, Multimed. Tools Appl. **79**, 289 (2020)
9. G. Tanaka, T. Yamane, J.B. Héroux, R. Nakane, N. Kanazawa, S. Takeda, H. Numata, D. Nakano, A. Hirose, Neural Netw. **115**, 100–123 (2019)
10. A. Morán, V. Canals, F.G.-Prado, C.F. Frasser, D. Radhakrishnan, S. Safavi, J.L. Rosselló, Cogn Comput. **15**, 1461–1469 (2023)
11. S.K.-Filipovic, P. Toliver, W. Johnson, R. Miller (2021). arXiv:2104.00751
12. M.F. Bear, B.W. Connors, M.A. Paradiso, *Neuroscience; Exploring the Brain*, 3rd edn. (Lippincott Williams & Wilkins Inc)
13. *Neuroscience: Exploring the Brain*, 3rd edn. (Lippincott Williams & Wilkins Publishers)
14. E.R. Kandel, in *Nerve Cells and Behavior*. ed. by E.R. Kandel, J.H. Schwartz, T.M. Jessell Principles of neural selection, 3rd ed. (Elsevier, New York, 1991)
15. M. Elia, *Energy Metabolism, Tissue Determinants and Cellular Corollaries* 61 (1992)
16. A. Gautam, T. Kohno, J. Robot. Netw. Artif. Life **7**, 22 (2020)
17. K. Yamazaki, V.-K. Vo-Ho, D. Bulsara, N. Le, Spiking neural networks and their applications: a review. Brain Sci. **12**, 863 (2022)
18. Y. Wei, F. Yao, Y. Kang, in 2024 3rd International Conference on Data Analytics, Computing and Artificial Intelligence (ICDACAI), Zakopane, Poland, 10 (2024)
19. M. Sorbaro, Q. Liu, M. Bortone, S. Sheik, Front. Neurosci. **14**, 662 (2020)
20. A.R. Omondi, J.C. Rajapakse, *FGAP Implementations of Neural Networks* (Springer)
21. https://www.amd.com/en/products/software/adaptive-socs-and-fpgas/vivado.html
22. https://www.intel.com/content/www/us/en/products/details/fpga/development-tools/quartus-prime.html
23. https://www.amd.com/en/products/software/adaptive-socs-and-fpgas/vitis/vitis-ide.html
24. https://www.ni.com
25. https://www.ni.com/docs/en-US/bundle/pxie-7862-specs
26. https://www.ni.com/docs/en-US/bundle/pxie-7868-specs
27. https://www.ni.com/docs/en-US/bundle/pxie-7915-specs
28. https://docs.amd.com/v/u/en-US/7-series-product-selection-guide
29. https://docs.amd.com/v/u/en-US/ultrascale-fpga-product-selection-guide
30. https://docs.amd.com/v/u/en-US/zynq-ultrascale-plus-product-selection-guide
31. H. Jaeger, German National Research Center for Information Technology GMD Technical Report **148**, 13 (2001)
32. H. Jaeger, Scholarpedia **2**, 2330 (2007)

33. W. Gerstner, Neuronal Dynamics: From Single Neurons To Networks And Models Of Cognition, Cambridge University Press (https://neuronaldynamics.epfl.ch/online/Ch1.S3.html).
34. W. Gerstner, *Neuronal Dynamics: From Single Neurons To Networks And Models Of Cognition* (Cambridge University Press). https://neuronaldynamics.epfl.ch/online/Ch3.S1.html
35. S. Haykin, *Adaptive Filter Theory*, 3rd edn, (Prentice-Hall, Inc)
36. H. Inoue, H. Tamura, A. Kitoh, X. Chen, Z. Byambadorj, T. Yajima, Y. Hotta, T. Iizuka, G. Tanaka, I.H. Inoue, *VLSI Technology and Circuits* 1 (2023)
37. H. Inoue, H. Tamura, A. Kitoh, X. Chen, Z. Byambadorj, T. Yajima, Y. Hotta, T. Iizuka, G. Tanaka, I.H. Inoue, Adv. Mater. **37**, 2407326 (2025)
38. E.M. Izhikevich, J.A. Gally, G.M. Edelman, Cereb. Cortex **14**, 933 (2004)
39. E.M. Izhikevich, IEEE Trans. Neural Netw. **15**, 1569 (2004)
40. K. Nakajima, Jpn. J. Appl. Phys. **59**, 060501 (2020)

Slow Electronics and Attractor

Kantaro Fujiwara

Abstract An attractor is an important concept in dynamical systems theory. In a dynamical system, an attractor is a set of states or points in phase space that the system tends to evolve toward as time progresses. In many real systems, the presence of nonlinearity introduces complex behavior, often leading to chaotic dynamics. Despite their unpredictable appearance, these chaotic systems are still governed by deterministic rules and are typically confined to a strange attractor. Understanding the attractor helps in analyzing the system's long-term behavior, even in the presence of chaotic fluctuations. These chaotic fluctuations, while appearing random, follow an underlying pattern that can be modeled and controlled. One of the central objectives of chaos control is to stabilize the system at a desired attractor. This may involve transitioning the system from chaotic dynamics to a more predictable state, such as a periodic orbit or a stable equilibrium. In this chapter, I will introduce how the dynamics of dysfunctional pancreatic β cells can be modeled and discuss a control method that utilizes on-demand electrical stimulation based on the principles of chaos control. Numerical simulation results demonstrate that the cell dynamics can be successfully regulated through a feedback mechanism, guiding the system back to a desired attractor despite initial dysfunction. These findings suggest that chaos control may enhance insulin secretion, potentially contributing to the treatment of type 2 diabetes. Our approach involves applying small electrical stimuli to activate the remaining healthy functions in dysfunctional cells. This method leverages homeostasis, a biological phenomenon characterized by long time constants. By applying weak, controlled stimuli, we can tap into the cell's inherent ability to restore balance. As a result, the concept of Slow Electronics can play a crucial role in realizing this treatment method, offering a novel way to manipulate biological systems for therapeutic purposes.

K. Fujiwara (✉)
Graduate School of Medicine, The University of Tokyo, Tokyo, Japan
e-mail: kantaro@g.ecc.u-tokyo.ac.jp

I. H. Inoue (ed.), *Slow Electronics with Reservoir Computing*,
https://doi.org/10.1007/978-981-96-8383-3_7

1 Attractor

An attractor is a concept in dynamical systems theory that describes a state or set of states toward which a system evolves over time. The word "attractor" is derived from "attract," meaning "to draw toward." Roughly speaking, it is a region that has the property of attracting surrounding orbits. Even if the system starts from different initial conditions, it tends to move toward the attractor. Attractors help explain stability and long-term behavior in various real systems. For example, in physics, attractors help in understanding phase transitions and fluid dynamics. In biology, they assist in modeling neural activity, genetic regulation, population dynamics, and more. In economics, attractors are used to predict market trends and equilibrium states.

Here we define the term attractor. Let $f : X \to X$ be a function describing the evolution of a dynamical system in a state space X. A subset $A \subset X$ is an attractor if the following properties hold:

- Attracting Property: There exists a neighborhood U of A such that for all $x \in U$, the trajectory of x under f converges to A:

$$\lim_{t\to\infty} d(f^t(x), A) = 0, \quad \forall x \in U$$

  where $d(x, A)$ is the distance from x to A, and $f^t(x)$ represents the system's state after t steps.
- Invariance: If a point is in the attractor, its trajectory remains within it:

$$f(A) = A$$

Attractors can take different forms depending on the complexity of the system. The first form of the attractor is, point attractor. This is called as fixed point, in different word. In the point attractor, a system settles into a single stable state. A pendulum in the presence of friction eventually stopping at its lowest point is an example of this kind of attractor.

The second form of the attractor is, limit cycle attractor. In the limit cycle attractor, a system follows a repeating cycle. The beating of a heart or the oscillation of a clock pendulum can be examples of this kind of attractor.

The third form of the attractor is, torus attractor. In the torus attractor, the system follows a quasi-periodic behavior. In this quasi-periodic behavior, the system is not strictly periodic, but is stable. Some types of coupled oscillators, such as certain planetary orbits can be an example of this kind of attractor.

The final form of the attractor is, strange attractor. This is called as chaotic attractor, in different word. In the strange attractor, the system follows a highly sensitive and unpredictable orbit, never repeating exactly but remaining within a constrained region. Weather patterns, turbulence in fluids are the examples of this kind of attractor.

The logistic map is a simple yet famous example of a dynamical system:

$$x_{t+1} = r x_t (1 - x_t)$$

For small r (e.g., $r = 2$), the system has a fixed-point attractor. On the other hand, for higher r (e.g., $r = 3.5$), the system exhibits periodic cycles. For very high r (e.g., $r = 4$), the system enters chaos, forming a strange attractor.

2 Nonlinear Dynamical Systems and Control

Real systems often have nonlinearity. This nonlinearity often leads to a phenomenon called chaos. Chaos is characterized by orbital instability and unpredictability. Orbital instability is famously known as the "butterfly effect," in which the behavior of a system changes drastically when a small perturbation is applied. Many of you may have heard examples, such as a single butterfly flapping its wings causing a typhoon. This is due to orbital instability, where a small error can amplify over time, resulting in vastly different trajectories (Fig. 1).

Unpredictability literally means that it is difficult to predict future states. As seen from the orbital instability described earlier, it is challenging to accurately predict the future of a chaotic system because a small error can grow into a significant one. These characteristics of chaos, such as orbital instability and unpredictability, can be problematic in many situations. In engineering, for example, such properties are considered drawbacks, since it is generally important for trajectories to be stable and predictable. Therefore, chaos control, i.e., how to control chaos, has become an important branch of control theory and is studied in various fields such as mathematics, physics, engineering, and life sciences [1]. If we can control chaos, we can gain predictability by stabilizing irregular behavior.

The main goal of control theory is to provide inputs to a system to drive it to a desired state. In chaos control, small perturbations are used to modify a chaotic system and bring it to a desired state. Chaos control theory has become a significant research topic focused on transforming irregular fluctuations, such as chaos, into

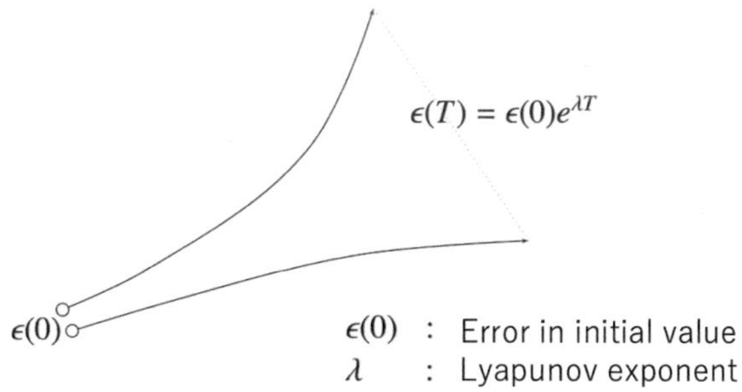

Fig. 1 Orbital instability in chaotic systems. The error $\epsilon(0)$ in the initial value increases exponentially with time

periodic fluctuations. There are various methods for chaos control, all of which are fundamentally based on stabilizing an unstable periodic orbit embedded in a chaotic attractor to convert chaotic fluctuations into periodic ones.

Moreover, once an orbit is attracted to such a region, it remains within that region. A periodic solution of a dynamical system that is locally unstable, i.e., unstable to small perturbations, is called an unstable periodic solution, and the orbit it forms is known as an unstable periodic orbit. It is well-known that a chaotic attractor contains numerous unstable periodic orbits embedded within it. Chaos control aims to manipulate these unstable periodic orbits and transform them into stable periodic orbits.

Research on chaos control began in the late 1980s. It focuses on stabilizing or guiding chaotic systems toward desired behaviors using minimal external interventions. Here, we introduce a typical chaos control method called OGY control. OGY control, which made the term "chaos control" famous, was proposed by Ott, Grebogi and Yorke in 1990 [2]. Technically speaking, OGY control is a technique to stabilize chaotic trajectories by applying small perturbations to system parameters. More simply, it causes a chaotic system to be controlled into a periodic system by adding small inputs to shift the system slightly. The advantage of OGY control is that the control inputs are small and can be controlled in a short period of time without changing the structure of the system. Because it is a very simple control theory and easy to handle, its application to many systems and extensions of OGY control have been studied.

For example, consider a continuous dynamical system in 3-dimensional space, which is written as follows, including the system parameter p.

$$\frac{dx}{dt} = F(x, p)$$

Let us assume that the equations describing the system are unknown and that the time series $x(t)$ is observed. Then we have the following.

$$X(t) = (x(t), x(t-T), x(t-2T), ..., x(t-mT))$$

This is called Takens embedding theorem [3], which means that when determining whether the original observed data is chaotic or not, it is possible to determine whether the system (= attractor) mapped (transformed) from the observed data to m-dimensional space behaves chaotically or not.

If d is the dimension of the original observed system, m should be

$$m \geq 2d + 1.$$

We consider using this $X(t)$ to stabilize the trajectory. An example for a 2-dimensional attractor is shown in Fig. 2.

Since chaos generally involves both periodic and irregular fluctuations, an attractor is described as a set of curves rather than a single curve. This consists of line segments

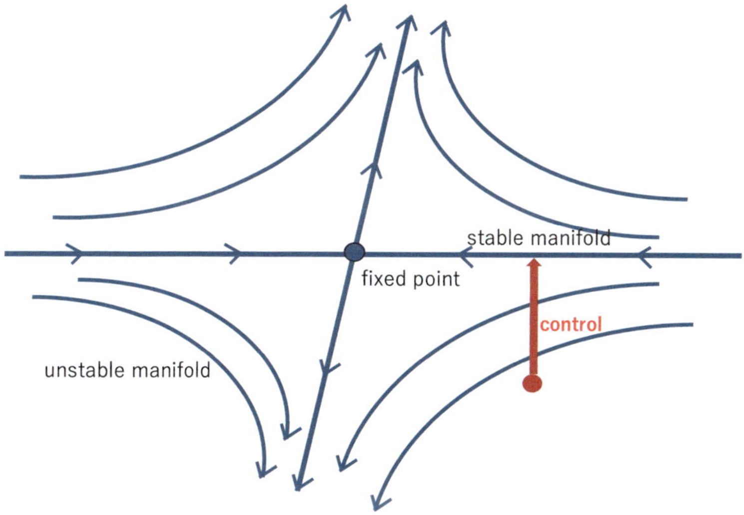

Fig. 2 Stable and unstable manifolds. OGY control is to consider controlling the transition from a point on an unstable manifold to a point on a stable manifold

that move in the direction of convergence to the fixed point, as shown in Fig. 2, while others move away from it, or move closer to it, but ultimately do not reach it. An attractor that converges to a fixed point is called a stable manifold, while a line segment that does not reach the fixed point or moves away from it is called an unstable manifold.

Chaos control is a method of convergence to the immovable point by applying an external force when approaching a neighborhood of the immovable point. Suppose the current value is at the red point in Fig. 2. In the next step, if we perturb the transition on the stable manifold, we have stabilized it.

For simplicity, consider a two-dimensional nonlinear discrete-time system with inputs

$$x(t+1) = f(x(t), I(t))$$

Let $x(t)$ be the state of the system at time t and $I(t)$ be the control input. When $I(t) = 0$, let $x_i(i = 1, 2, 3, \ldots., T)$. However,

$$x_{i+1} = f(x_i, 0) \quad i = 1, 2, 3, \ldots, T-1$$

$$x_1 = f(x_T, 0)$$

The controller applies an input when the difference between the system state $x(t)$ and the target trajectory x_i is sufficiently small, otherwise it does not apply an input. Thus, the output of the controller is

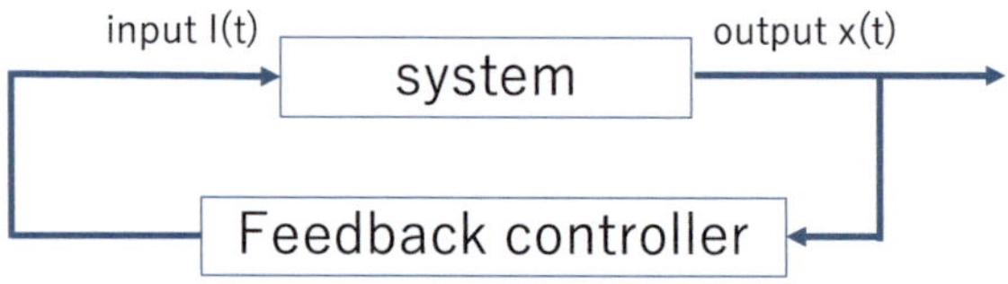

Fig. 3 Block diagram of OGY control. The feedback input $I(t)$ is applied to the system from the feedback controller by the value of the output $x(t)$ from the chaotic system

$$I(t+1) = \begin{cases} K_{a(t)}(x_{a(t)} - x(t)) \text{ if } \|x_{a(t)} - x(t)\| < \epsilon \\ 0 \qquad\qquad\qquad\qquad \text{otherwise} \end{cases}$$

$K_i (i = 1, ..., T)$ represents the amplitude of the feedback when the state is in the neighborhood of the periodic point X_i. In OGY control, K_i is set so that the state approaches the stable manifold of the periodic point. A conceptual diagram of OGY control is shown in Fig. 3.

Assuming that the value of the output $x(t)$ from the chaotic system is observable, the feedback input $I(t)$ is applied to the system from the feedback controller by that value. Such feedback control is the basic idea of OGY control.

In the next section, we will introduce a theoretical study that applies such an idea of OGY control to the control of β cells in the pancreas.

3 Therapy Using a Pancreatic β Cell Model

The effectiveness of chaos control on biological systems has been demonstrated in cardiac muscle tissue in rabbits [4] and the hippocampus in rats [5]. Specifically, artificially induced arrhythmia has been controlled successfully to a regular beat in a rabbit heart, with applicability to remedy for heart disease in mind. Aiming at medical application, it could be possible to select the optimum stimulation voltages by using chaotic features of tremor in parkinsonian patients [6].

In this section, we introduce the example of application of chaos control to diabetes. Diabetes is one of the largest diseases in human history with more than 400 million patients worldwide [7]. In Japan, the number of people with type 2 diabetes is said to be nearly 20 million, including those with pre-diabetes, and the number is still increasing [8].

Currently, two types of drug therapy are considered effective for type 2 diabetes: insulin therapy and oral hypoglycemic agents. Both insulin therapy and oral hypoglycemic agents impose a heavy burden on patients and have serious side effects. Therefore, there is a growing need for alternative treatments that can improve the quality of life (QOL) of patients.

At present, the main alternative treatments for diabetes in development are regenerative medicine approaches that attempt to regenerate and proliferate dysfunctional cells and genotype-based therapies, but further progress in stem cell research is essential, and this requires a great deal of time and effort. On the other hand, one treatment that may solve these problems is electrical stimulation therapy. This is a new technique that stimulates insulin secretion from pancreatic β cells through electrical or magnetic stimulation instead of drugs.

Pancreatic β cells are located in islets of Langerhans in the pancreas and secrete insulin, a hormone that is responsible for regulating blood glucose levels. Islets of Langerhans (pancreatic islets) are endocrine tissues scattered like islands inside the pancreas, and there are about 1–2 million of them in the entire pancreas. Islets are a collection of endocrine cells and are composed of α, β, and γ cells. Among them, β cells constitute the majority, accounting for 75–80% of the total.

The mechanism of insulin secretion in pancreatic β cells has been gradually becoming clearer in recent years. Insulin secretion in pancreatic β cells is caused by an increase in intracellular calcium concentration. Glucose initially enters the cell and is metabolized, resulting in an increase in adenosine triphosphate (ATP) at the expense of adenosine diphosphate (ADP). This increases the ATP/ADP ratio and closes the ATP-sensitive potassium channel (KATP). This closure causes depolarization, a rapid increase in membrane potential, which opens voltage-dependent calcium channels (VDCCs), allowing calcium ions to enter the cell and triggering insulin secretion. As described above, pancreatic β cells cause insulin secretion in association with changes in membrane potential, and insulin secretion and membrane potential changes are thought to be related. For example, it is well known that pancreatic β cells secrete insulin in response to bursts of membrane potential.

The activity of pancreatic β cells in diabetic patients shows various characteristics. First, some experimental results suggest that the frequency of membrane potential bursts is reduced in pancreatic β cells of diabetic patients [9, 10]. Also, results suggesting that bursting activity of membrane potentials becomes irregular have been reported [11–13]. These results suggest that increasing the burst frequency or regularity of membrane potential bursts in pancreatic β cells stimulates insulin secretion. However, it is important to maintain an appropriate burst frequency, because an excessively high burst frequency may lead to hypoglycemia and pancreatic hypofunction due to insulin hypersecretion. In type 2 diabetes, noninvasive electrical or magnetic stimulation is thought to be effective in restoring the function of pancreatic β cells, and in principle, it is possible to control insulin secretion by controlling these cells. It is based on the simple and clear principle of restoring the function of pancreatic β cells by applying artificial stimulation such as electricity or magnetism to these cells. This possibility has been suggested from a medical viewpoint, as in the case of the restoration of function by electrical stimulation of stomach wall cells.

So what kind of electrical stimulation is effective as a treatment? To find out, mathematical modeling is needed. In this chapter, we introduce the electrical stimulation therapy for type 2 diabetes using the study by Koyama et al. as an example [14].

First, the Chay model [15] proposed by Chay is used as a mathematical model to reproduce burst activity of pancreatic β cells. This model is as follows.

$$\frac{dV}{dt} = g_I^* m_\infty^3 h_\infty (V_I - V) + g_{K,V}^* n^4 (V_K - V)$$
$$+ g_{K,C}^* \frac{C}{1+C} (V_K - V) + g_L^* (V_L - V) + I_{\text{input}} \tag{1}$$
$$\frac{dn}{dt} = \frac{(n_\infty - n)}{\tau_n} \tag{2}$$
$$\frac{dC}{dt} = \rho \left[m_\infty^3 h_\infty (V_C - V) - k_C C \right] \tag{3}$$

Here, V, n, C represent membrane potential, open probability of the voltage-dependent K^+ channel, intra-cell Ca^{2+} ion concentration divided by dissociation constant from receptor. V_I, V_K, V_C, V_L represent reversal potential of the mixed Na^+ and Ca^{2+} ions, K^+ ion, Ca^{2+} ion, and leak ion. g_I^*, $g_{K,V}^*$, $g_{K,C}^*$, g_L^* represent maximum conductance of the mixed Na^+ and Ca^{2+} ions, voltage dependent K^+, Ca^{2+} dependent K^+, leak ion. The probabilities of activation and inactivation of the mixed channel are m_∞ and h_∞, respectively. n_∞ is the steady-state value of n. Also, τ_n is the relaxation time of the potential gate K^+ channel, k_C is the outflow rate constant of intracellular Ca^{2+}, ρ is the proportionality constant, and I_{input} is the stimulus current.

Here, m_∞, h_∞ and n_∞ are replaced by the steady state described by $y_\infty = \frac{\alpha_y}{\alpha_y} + \beta_y$.

$$\alpha_m = \frac{0.1(25+V)}{1-\exp(-0.1V-2.5)}$$
$$\beta_m = 4\exp(-(V+50)/18)$$
$$\alpha_h = 0.07\exp(-0.05V-2.5)$$
$$\beta_h = \frac{1}{1+\exp(-0.1V-2)}$$
$$\alpha_n = \frac{0.01(20+V)}{1-\exp(-0.1V-2)}$$
$$\beta_n = 0.125\exp(-(V+30)/80)$$

τ_n is defined as follows.

$$\tau_n = \left[230\,(\alpha_n + \beta_n)\right]^{-1}$$

The Chay model is a mathematical model that shows periodic behavior depending on the parameters, but chaotic and irregular behavior when the values of the parameters are changed. It is a mathematical model that can reproduce various behaviors and is often used as a mathematical model of pancreatic β cells.

It is inferred from experimental results that burst activity patterns of dysfunctional pancreatic β cells show chaotic and irregular behaviors. On the other hand, burst activity patterns of healthy pancreatic β cells are known to be regular.

Pancreatic β cell status	Activity pattern of the membrane potential
Healthy	Regular
Dysfunctional	Irregular

As shown in the table above, dysfunctional pancreatic β cells and healthy pancreatic β cells have different activity patterns. Based on the experimental results, the mathematical model reproduced the dysfunctional cells by setting the parameters of the Chay model to values that exhibit chaotic dynamics. The Chay model exhibits nonlinear dynamics and deterministic chaos in some parameter regions. The return of a dysfunctional pancreatic β cell to the original healthy pancreatic β cell corresponds to a transition of the Chay model from a chaotic state to a periodic state in the mathematical model. Control theory is used for such state transitions.

Here, we will consider what type of electrical signal is used for control. We consider it as an external input stimulus that provides input according to the state of the membrane potential. The input $I(t)$ is shown below.

$$I_{\text{input}}(t) = \begin{cases} a & \left(t_{s_i} \le t \le t_{s_i} + l\right) \\ 0 & (\text{otherwise}) \end{cases} \tag{4}$$

In this equation, a represents the input current value (stimulus intensity) and l represents the stimulus input time (stimulus width). Also, t_{s_i} represents the time for the cell membrane potential $V(t)$ to reach the set membrane potential value V_{th} corresponding to the input threshold at the ith time ($i = 1, 2, \ldots$). This decision is made only when $\frac{dV(t)}{dt} > 0$, in consideration of input efficiency, to limit unwanted input. In other words, current with intensity a and stimulus width l is input from the moment the membrane potential $V(t)$ reaches V_{th} with $dV(t)/dt > 0$. Each time the cell's membrane potential reaches the set value V_{th}, the input of this stimulus is turned on. After a set time l has elapsed, the input is turned off again. Therefore, this is a digital current as it is a pulsed current that is repeatedly turned on and off. A schematic diagram of the control scheme is shown in the following Fig. 4 (Fig. 5).

The membrane potential $V(t)$ output from the pancreatic β cells is always measured and the feedback input $I(t)$ from the feedback controller is applied to the pancreatic β cells according to Equation (7). These feedback inputs are the control inputs. You may have noticed that this is the OGY control we did in the previous section. The activity of the pancreatic β cells is controlled by the OGY controller. Now, let us see how the irregular firing pattern is controlled to a regular firing pattern by such a control input. Figure 6 shows the result of applying feedback control to the chaotically irregular activity of pancreatic β cells.

As shown in Fig. 6, the originally irregular firing pattern became a regular firing pattern as a result of feedback control. In other words, we can see that what was dysfunctional has become healthy. Although this result is only a fraction of the

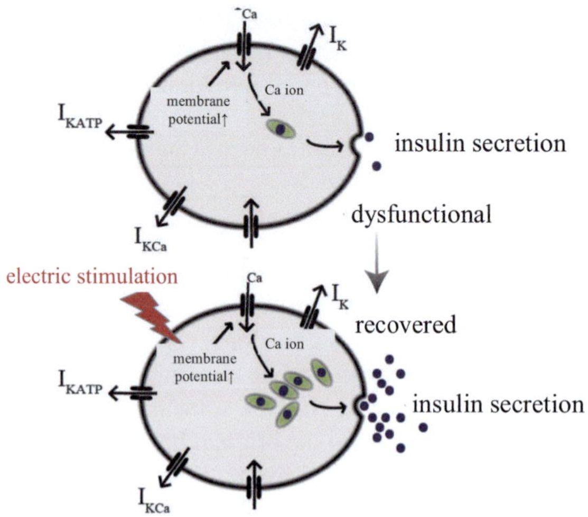

Fig. 4 Schematic diagram of pancreatic β cells. Function is recovered by external weak electrical stimulation

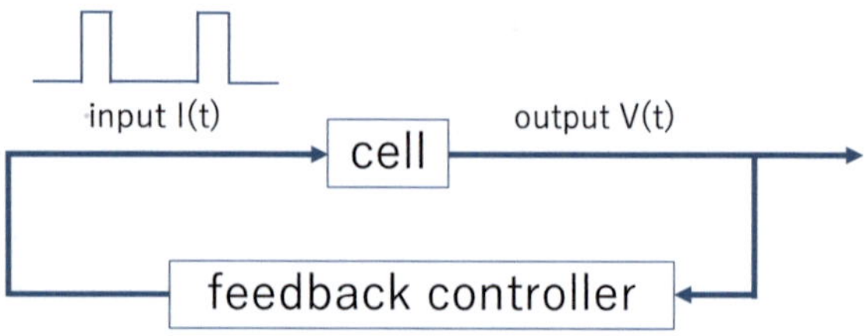

Fig. 5 Block diagram of on-demand feedback control. The feedback input $I(t)$ is applied from the feedback controller to the cell according to the value of the output $V(t)$ from the cell

results of the numerical simulation, such control proved to be effective for a variety of irregular activity patterns.

Also noteworthy is that this control does not force pancreatic β cells to behave in a periodic manner, but rather makes them behave in a periodic manner at a natural frequency. Thus, the strength of the control input is very weak. This is quite different from the way a cardiac pacemaker works. This controller assists spontaneous and regular cellular activity, rather than forcibly and continuously stimulating the cells to beat a rhythm from the outside like a cardiac pacemaker.

See the Fig. 7. This figure superimposes a trajectory that was originally normal and a trajectory that has become dysfunctional and is now controlled. From this figure, we can see that the controlled and recovered trajectory is almost the same as the original normal trajectory. What is noteworthy here is that this trajectory is originally the attractor of the cell itself. What was displaced from the attractor due to dysfunction was brought back to the attractor by a weak control input.

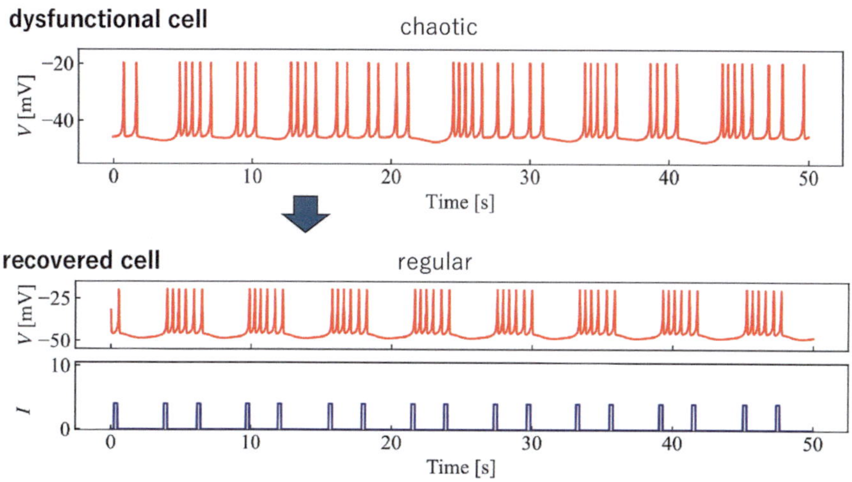

Fig. 6 The membrane potential of the pancreatic β cell is shown in red and the control input is shown in blue. The chaotic activity of dysfunctional pancreatic β cells was recovered to regular activity as a result of feedback control

Fig. 7 The trajectory when it was originally normal is superimposed on the trajectory when it became dysfunctional and controlled. We can see that the controlled trajectory is almost identical to the original normal trajectory

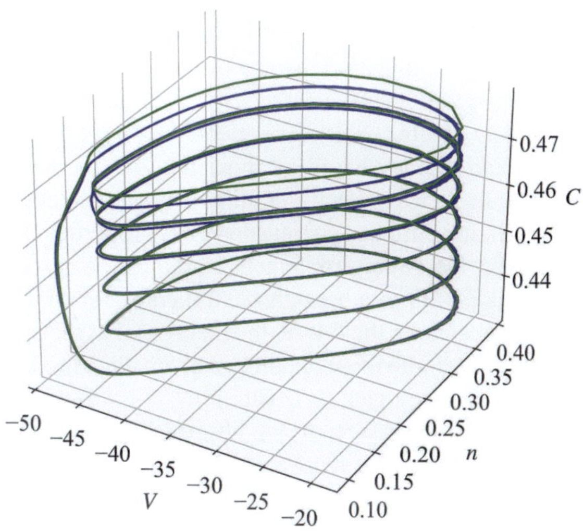

Our therapy method's characteristic is that, the controller acts more like an occasional hand reaching out to recover the original function rather than a constant helping hand, so to speak. In the sense that only a weak control current is required, it is also possible to keep the side effects on the human body small, and in terms of device implementation, it is a treatment method that is easy to put into practical use.

Furthermore, the proposed method assumes electrical activation of pancreatic β cells, rather than chemical treatment that externally alters cell characteristics. In other words, the target cells are not those that are completely dead, but those that have become dysfunctional but can be recovered with a little tinkering. Incidentally, this is not an effective treatment for cells that are completely dead. In addition, electrical stimulation is expected to spontaneously improve the function of pancreatic β cells because it is an external control input that helps the cells return to a healthy state. In the sense of evoking the original functions, this type of controlled therapy is distinct from conventional therapies. It is important to note that there is both a therapeutic aspect and a rehabilitation aspect. This is because the weak electrical stimulation therapy stimulates the cells' natural functions, enabling them to function normally without electrical stimulation. In general drug therapy such as insulin injections, continuous external injection of insulin may cause cells to mistakenly believe that they do not need to secrete insulin, or they may stop secreting insulin even more than before. This is the downside of "drugging," so to speak, and because of this reliance on drugs, ironically, cellular function is lost. The original treatment is to recover the body to normalcy. In this sense, electrical stimulation therapy is considered to be the original treatment.

Therapy	Effect	Advantage	Disadvantage
Drug	reduce blood glucose	reliability	side effects
Electrical stimulation	enhance insulin secretion	small side effects	poor clinical trials

Electrical stimulation therapy also reduces the physical burden on the patient by using energy-efficient, weak electrical stimulation. Thus, it is expected to improve the quality of life (QOL) for patients with type 2 diabetes. The proposed electrical stimulation therapy based on this study may be incorporated into future electrical devices. In fact, an electronic circuit implementation of neuronal chaotic control, incorporating similar dynamical system properties, has already been proposed [16]. Therefore, the circuit implementation of our control method is expected to lead to the development of a breakthrough therapy for controlling electrical stimulation in type 2 diabetes. In the future, the theoretical aspects of implementing such a control device will require more detailed mathematical modeling and numerical simulations to carefully investigate its feasibility.

In the previous section, we observed the successful control of a dysfunctional cell transitioning to a recovered cell. This is a good example of how to effectively use the concept of an attractor and how to successfully control and lead a system from an off-attractor state to an attractor. Successful control from an off-attractor state to an attractor requires strategic interventions: small, well-timed adjustments that nudge the system along stable trajectories, rather than forcing abrupt shifts. In biological systems, homeostasis ensures stability by regulating internal conditions, while attractors in dynamical models represent the stable states toward which the system naturally evolves. Life maintains homeostasis, and the same applies to the pancreas and its β cells. Even if the body deviates slightly from the normal state due to illness, homeostasis allows it to return to the normal state. Of course, homeostasis will not work if the cells die. It is important to note that it is dysfunction, not death, that is key. A state of dysfunction is a condition in which cells can still effectively utilize homeostasis. This concept aligns perfectly with the idea of slow electronics. Slow electronics is a new field of research that utilizes the properties of systems with long time constants. While its applications are still unclear, the control theory of β cells introduced here can be considered a good example of its potential application.

References

1. S. Boccaletti, C. Grebogi, Y.C. Lai, H. Mancini, D. Maza, Phys. Rep. **329**(3), 103 (2000)
2. E. Ott, C. Grebogi, J.A. Yorke, Phys. Rev. Lett. **64**, 1196 (1990)
3. F. Takens, in *Dynamical Systems and Turbulence, Warwick 1980* (Springer, Berlin Heidelberg, Berlin, Heidelberg, 1981), pp. 366–381
4. A. Garfinkel, M.L. Spano, W.L. Ditto, J.N. Weiss, Science **257**(5073), 1230 (1992)
5. S.J. Schiff, K. Jerger, D.H. Duong, T. Chang, M.L. Spano, W.L. Ditto, Nature **370**, 615 (1994)
6. M. SadeghiRazlighi, A.H. Jafari, S.M. Firoozabadi, G.A. Shahidi, Australasian Phys. Eng. Sci. Med. **35**, 25 (2012)
7. H. Sun, P. Saeedi, S. Karuranga, M. Pinkepank, K. Ogurtsova, B.B. Duncan, C. Stein, A. Basit, J.C. Chan, J.C. Mbanya, M.E. Pavkov, A. Ramachandaran, S.H. Wild, S. James, W.H. Herman, P. Zhang, C. Bommer, S. Kuo, E.J. Boyko, D.J. Magliano, Diabetes Res. Clin. Pract. **183**, 109119 (2022)
8. N. Kawakami, S. Araki, M. Kawashima, T. Kawaguchi, Diabetol. Int. **9**(3), 232 (2018)
9. F. Martin, B. Soria, Cell Calcium **20**(5), 409 (1996)
10. J. Campbell, C. Newgard, Nat. Rev. Mol. Cell Biol. **22**(3), 142 (2021)
11. T. Kanno, T. Miyano, I. Tokuda, J. Galvanovskis, M. Wakui, Phys. D **226**, 107 (2007)
12. M.S. Klemen, J. Dolensek, M.S. Rupnik, A. Stozer, Islets **9**, 109 (2017)
13. P. Rorsman, F.M. Ashcroft, Physiol. Rev. **98**(1), 117 (2018)
14. K. Koyama, H. Ando, K. Fujiwara, AIP Adv. **13**(4), 045317 (2023)
15. T.R. Chay, Phys. D **16**, 233 (1985)
16. H. Ando, A. Nakano, Y. Horio, K. Aihara, in *2009 IEEE International Symposium on Circuits and Systems* (IEEE, 2009), pp. 2621–2624

Decoding the Unseen, Shaping the Future

Isao H. Inoue

Abstract In this chapter, we summarise the main developments of slow electronics and their impact, and examine future possibilities in human-centric applications. Technological innovations–such as neuromorphic devices with long time constants, reservoir circuits, and algorithms inspired by the principle of attractors–have brought about "slow electronics," which makes it possible to process, in real time and with ultra-low power consumption, those time-series data with extremely long characteristic timescales that were previously difficult to handle using conventional ultra-fast digital control. Slow electronics is expected to be adopted into our daily lives–from autonomous regulation of physiological states and everyday activities to the enhancement of specialised skills–and to evolve into indispensable edge devices for a human-centric life, supporting quality of life improvements, emotional care, and cognitive resilience. In this chapter, by bridging technical and human/social perspectives to contemplate "Where is slow electronics headed?" and by looking ahead to its long-term contributions to technology and society, we bring this book to a close.

1 Introduction

Until now, in the world of electronics engineering, progress has largely been driven by the theme of "faster and higher performance." However, **slow electronics** proposes a different approach. This is a new paradigm of electronics engineering that deliberately operates devices "slowly" to achieve harmony with humans and the environment. For example, drawing inspiration from how biological neural networks process signals over long timescales with ultra-low power, slow electronics emphasises devices and computing mechanisms that handle **time-series data with extremely long characteristic timescales**. Now, slow electronics is gaining attention because, as AI and IoT become ubiquitous, the limits and downsides of constantly processing huge volumes of data at high speed have become apparent. High-speed computing

I. H. Inoue (✉)
National Institute of Advanced Industrial Science and Technology (AIST), Tsukuba, Japan
e-mail: isaocaius@gmail.com

I. H. Inoue (ed.), *Slow Electronics with Reservoir Computing*,
https://doi.org/10.1007/978-981-96-8383-3_8

is certainly convenient, but it comes with high energy consumption and can be out of sync with the characteristic timescales of phenomena around humans. Slow electronics is, therefore, attracting attention as an electronics technology that operates in harmony with human society and the environment.

Slow electronics differs greatly from conventional electronics in its **design philosophy**. Traditionally, electronics has pursued switching speeds on the order of nanoseconds and clock frequencies in the GHz range. In contrast, slow electronics actively utilises devices with **long time constants** on the order of milliseconds, seconds, or even longer. By using components whose state changes slowly–for example, devices in which ions play a role–once a state change occurs, its effect persists, enabling behaviours such as remembering the temporal context of inputs. This is analogous to how biological neurons integrate signals gradually or adapt slowly to environmental changes. In other words, slow electronics can be seen as an attempt to incorporate the **fluctuations** and **flexibility** inherent in biological systems into electronic circuits. The resulting systems are endowed with **human affinity**–they are intuitive and easy for humans to interact with–while achieving energy-efficient, sustainable computation. This holds great potential for applications in edge computing (intelligent devices at the periphery that do not rely on the cloud), wearable AI devices, and other intelligent machines that assist humans.

In this chapter, we explore the future vision outlined by slow electronics and its societal impact. First, we briefly revisit the technical highlights covered in Chaps. 1–7 to reaffirm the foundation of slow electronics. Building on that, we look ahead at what technological developments can be expected in the next 15 years, 25 years, and beyond, considering the implications for society and potential new lifestyles. Furthermore, we discuss specific application examples of utilising slow electronics as **human-centric devices**–in areas such as medicine/healthcare, welfare/caregiving, education/learning, and wearable daily devices–from the perspective of a technology that evolves together with humans. Finally, taking all of these into account, we illustrate a day in the life of a future family realised by slow electronics, bringing this book to its conclusion.

2 What Have We Achieved So Far?

Before drawing the overall picture of slow electronics, let us review the main topics covered in Chaps. 1 through 7. The technical elements and principles learned throughout this book form the foundation for discussing future outlooks.

2.1 *What is Slow Electronics? (Chap. 1)*

Chapter 1 introduced the concept of "**slow electronics**." The need for an ultra-low-power computing architecture capable of responding to slow-changing phenomena–taking hints from the human brain–was discussed, and slow electronics was proposed as a new framework for **real-time processing of long-timescale data**. Specifically, the chapter pointed out the challenges that conventional artificial neural networks (ANNs) face in terms of power consumption and handling slow events, and suggested solutions such as neuromorphic devices with long time constants, spiking neural networks (SNNs), and reservoir computing. The three pillars of slow electronics (long time-constant devices, environment-responsive architectures, and attractor-based algorithms) together enable human-friendly, ultra-low-power computation, hinting at a edge computing that does not rely on the cloud.

2.2 *Reservoir Computing (Chap. 2)*

As one of the computational methods supporting slow electronics, **reservoir computing** (echo state networks, ESNs) was highlighted in Chap. 2. Reservoir computing is a lightweight machine learning approach well-suited for time-series pattern processing with low training cost. The chapter focused on the ability to handle different timescales, introducing models such as the leaky integrator ESN (LI-ESN), which can adjust its response speed to inputs, and the diverse timescale ESN (DTS-ESN), where each node has a different time constant. These models demonstrated that by controlling the speed of state changes in the reservoir (e.g. leak rate or distribution of time constants), one can successfully approximate multi-scale dynamic phenomena. In fact, it has been reported that reservoirs with a richer diversity of time constants exhibit higher performance on complex time-series prediction tasks [1]. This provides a computational foundation that aligns with **slow electronics' ethos of leveraging slow dynamics**.

2.3 *Core Devices for Slow Electronics (Chap. 3)*

Chapter 3 discussed material and device-level approaches for realising slow electronics. The focus was placed on the slow movement of ions such as oxygen defects [2] and protons [3], and devices employing such ion-conductive materials were introduced. By harnessing **sluggish physical phenomena in solids**–like resistance changes due to oxygen-defect migration or threshold shifts caused by proton motion–one can create devices whose state changes slowly over time. For example, whereas a typical MOSFET switches current within nanoseconds of a gate voltage being applied, the "slow FET" introduced in Chap. 3 takes time for the current to gradually ramp up, requiring a delay before reaching its steady value. In such devices, the effect of an applied signal remains within the device as a kind of temporal memory, yielding unique characteristics such as the ability to determine "when a signal was applied" after the fact. Chapter 3 also presented foundational work on slow-electronic devices using functional materials like ferroelectrics and phase-change materials, marking a first step toward the hardware realisation of slow electronics.

2.4 Hardware Neurons with Analog CMOS (Chap. 4)

Chapter 4 explored efforts to realise the principles of slow electronics using existing CMOS technology by implementing neuron circuits with analogue electronics. A leaky integrate-and-fire (LIF) artificial neuron circuit was designed and fabricated in a 65 nm CMOS process, achieving a long time constant of approximately 1.2 ms within an ultra-compact circuit area of 54.8×25.4 μm while consuming only a few microwatts. Traditional implementations of long time constants required large capacitors or high resistance values, limiting scalability. This study introduced a **pseudo-resistor (PR)** technique using MOS transistors in a special configuration to achieve extremely high resistance in a compact form, enabling efficient integration into neuromorphic systems [4].

Beyond LIF neurons, Chap. 4 also introduced a **CMOS-based time-domain analogue neuron and synapse circuit**, designed for energy-efficient neuromorphic applications. Unlike conventional designs that rely on voltage levels, this circuit uses **time-domain encoding**, where neuronal activity is represented by signal transition intervals. The neuron module employs an **inverter-based comparator**, simplifying threshold detection, while the synapse module incorporates a **voltage-controlled ring oscillator** to generate frequency-modulated pulses. Synaptic weights are dynamically adjusted using a **delay-based mechanism**, offering an efficient alternative to conventional current-domain circuits [5].

Experimental results demonstrated that this approach enables **scalable and low-power neuromorphic processing**, maintaining biologically plausible neuronal dynamics. Compared to prior designs, the proposed circuit achieves superior area efficiency and significantly reduced power consumption. As demonstrated in Chap. 4, CMOS technology is well suited for **realising slow yet efficient neuromorphic circuits**, opening new possibilities for ultra-low-power, time-sensitive artificial intelligence systems.

2.5 Learning and Inference in Slow Electronics (Chap. 5)

Chapter 5 tackled the challenge of how to perform real-time learning and inference on systems built with slow devices, using time-series anomaly detection as a case study. Conventional methods had shortcomings: anomaly detection with a sliding window often suffers from delayed detection of changes, and deep learning-based methods can require excessive training time. To address this, Chap. 5 proposed the **Mahalanobis Distance of Reservoir States (MD-RS)** method [6], which combines reservoir computing's efficient temporal encoding with statistical anomaly scoring based on the Mahalanobis distance. In this approach, the distribution of features obtained from the reservoir's dynamic states is modelled, and anomalies are detected by measuring deviations from this distribution for new inputs. In evaluations on the UCR time-series anomaly detection benchmark, the proposed MD-RS consistently demonstrated superior timing accuracy compared to a sliding-window method (MD-SW) and a reconstruction error-based method (RC-SRE), managing to flag anomalies with virtually no delay. Additionally, the chapter introduced an application to **handwriting authentication (signature verification)** [2], suggesting that practical learning and inference are feasible even under the slow electronics paradigm.

2.6 Real-Time FPGA Implementation (Chap. 6)

As a proof of concept for slow electronics, Chap. 6 presented an example of a real-time system implemented on an FPGA. In this demonstration, a hardware reservoir system was built to learn and replay handwritten pen trajectory data in real time. The system processes input pen-stroke coordinates as time-series data, and implements **a reservoir circuit on the FPGA** consisting of on the order of tens to hundreds of spiking neurons with two-dimensional input and output nodes. By incorporating into this FPGA-based reservoir the anomaly detection algorithm from Chap. 5 and **the multi-timescale neurons introduced in Chap.** 2, it was shown that **real-world data can be processed and learned on the fly**. In the pilot experiment, the system continuously learned a pen-drawn trajectory and succeeded in drawing that trajectory predictively from its output layer. This result indicates that the core technologies of slow electronics can operate on existing digital hardware, suggesting that in the near future we might implement them in embedded devices or wearables.

2.7 Application to Biological Rhythms: Attractors and Medicine (Chap. 7)

As an example of applying **slow electronics concepts to the medical field**, Chap. 7 introduced research in which slow biological dynamics are manipulated to treat disease. Specifically, it discussed an attempt to improve symptoms of type 2 diabetes by applying weak electrical stimulation to the slow oscillatory activity (the bursting electrical rhythm and insulin secretion cycle) of pancreatic β-cells. Using Koyama et al.'s study as an example, the chapter analysed the effect of external stimulation on a pancreatic β-cell electrical activity model (the Chay model), examining how stimuli influence the cell's state (attractor) [7]. The results showed that low-frequency, weak electrical stimulation can restore an abnormal insulin secretion rhythm to a normal periodic cycle. A schematic diagram illustrated how, under stimulation, the patterns of membrane potential oscillation and intracellular Ca^{2+} concentration change, leading to recovery of insulin secretion. This example suggests that the **slow electronics approach (gently modulating a system's attractor state)** could be effective as a form of medical intervention. In the future, one can envision implantable devices that continuously monitor biological rhythms in real time and provide gentle stimulation as needed to prevent or treat disease–an application of slow electronics as a technology that truly stays close to the person.

In summary, throughout Chaps. 1–7 we have explored various elements that compose slow electronics. New principles (Chap. 2), device technologies (Chaps. 3 and 4), algorithms (Chap. 5), hardware implementations (Chap. 6), and potential applications (Chap. 7) were investigated, examining the potential of slow electronics from multiple angles. Now, building on these insights, let us look ahead to the future envisioned by slow electronics.

3 Applications as Human-Centric Devices

As concrete examples of slow electronics in use, its deployment in human-centric devices across medicine, welfare, education, and wearables is highly anticipated. In this section, we describe the possibilities that slow electronics may offer in each of these domains.

3.1 Applications in Medicine and Healthcare

Medicine and healthcare are particularly promising fields for slow electronics. In wearable devices that continuously monitor biosignals such as heart rate or EEG, slow electronics can significantly reduce power consumption, enabling extended monitoring without frequent recharging. This facilitates early detection of chronic conditions or post-surgery complications while allowing patients to maintain their normal routines. In implantable medical devices–such as pacemakers or deep-brain stimulators–slow electronics can dramatically reduce battery replacement frequency, lessening patient burden and mitigating surgical risks. Additionally, by enabling localised processing instead of cloud-based analysis, slow electronics eliminates concerns over data privacy breaches and cyberattacks. Since the device learns and processes data directly on the user's body, there is **no need to transmit sensitive medical information externally**, ensuring greater security.

A key advantage of slow electronics is its ability to learn and adapt to **an individual's unique physiological patterns**. Unlike conventional monitoring systems that rely on generalised thresholds, slow electronics can distinguish between normal variations for a specific user and subtle deviations that may indicate health issues. This enables truly personalised diagnostics, where the system minimises false alarms while promptly alerting the user or physician to genuine concerns. By making medical devices more autonomous, personalised, and secure, slow electronics has the potential to revolutionise healthcare, paving the way for a future where **highly individualised medical support** is seamlessly integrated into daily life.

3.2 Applications in Welfare and Caregiving

Slow electronics offers transformative potential in welfare and caregiving by enabling truly personalised support tools. For instance, in elderly care, a monitoring sensor that learns an individual's unique daily rhythms and subtle behavioural patterns–**their personal "homeostasis"**–can detect gradual changes such as declining activity levels or disrupted sleep. Unlike conventional systems, which rely on generic thresholds, slow electronics adapts to each user, minimising false alarms while identifying subtle warning signs that might otherwise go unnoticed. Since these ultra-energy-efficient devices require infrequent battery changes, they can function continuously once installed, reducing the burden on caregivers and family members. Furthermore, slow electronics eliminates reliance on cloud-based data processing, protecting privacy and preventing cyberattacks. By learning and processing information locally, the device remains intimately attuned to the user, much like **a dedicated personal care manager who understands their specific needs and habits**.

Additionally, slow electronics enhances assistive technologies such as communication aids and prosthetic limbs by making them lighter, more compact, and longer-lasting. A robotic prosthetic arm that learns a user's unique muscle signals and movement patterns could provide smoother, more intuitive control, reducing physical strain. Through these innovations, slow electronics fosters a more inclusive society, where **personalised, secure, and discreet care technologies** enable individuals to live with greater independence and peace of mind.

3.3 Applications in Education and Learning

In the field of education, slow electronics could be leveraged as a technology that stays attuned to each learner. If a system can learn and adapt to **an individual user's distinctive activity patterns**, it could achieve personalised optimisation that responds to long-term changes in understanding or interest. For example, imagine a device that quietly monitors each student's level of concentration or fatigue and, in coordination with adjusting environmental factors like room lighting or temperature, also modulates the pace of learning materials to support comfortable and efficient learning. By detecting changes in a learner's heart rate or facial expressions, the technology can **act behind the scenes as a gentle assistant**, subtly supporting the student. The result could be a new form of education that ultimately draws out students' innate creativity and autonomy.

3.4 Applications in Wearables and Daily Devices

Wearable devices are a key application for slow electronics, offering a level of personalised intelligence and security that conventional cloud-based systems cannot match. Unlike standard smartwatches or fitness trackers that rely on external servers, slow electronics processes data locally, eliminating the risks of privacy breaches and cyberattacks. This means users can enjoy the benefits of continuous health monitoring, activity tracking, and intelligent assistance **without exposing their personal data to third parties**. More importantly, these devices learn and adapt to the individual, functioning like **a dedicated personal secretary** that understands the user's unique routines, preferences, and needs. Rather than simply collecting raw data, slow electronics enables context-aware insights, where the system recognises meaningful patterns–whether it is subtle shifts in daily habits, stress levels, or productivity trends–and provides highly personalised support.

By removing dependence on cloud-based AI and ensuring that all processing happens on the device itself, slow electronics creates a future where wearable technology is not just convenient but also truly private, secure, and intimately attuned to each user.

4 Future Prospects for Slow Electronics

Slow electronics is still an emerging field of research, but it holds tremendous potential for the future. Envisioning how slow electronics will advance and what new technologies might emerge is truly exciting. Here, we sketch out a vision along a timeline: over the next 15 years, 25 years, and further into the future.

In the short term (roughly the **next 15 years**), the foundational technologies of slow electronics are expected to advance significantly toward practical realisation. Prototype devices and small-scale systems currently in the research stage will be further refined, achieving more stable and reliable operation. Insights gained from research projects will be implemented in prototype implementations, and initial application examples will start to appear one after another. At first, as demonstrated with the FPGA-based pilot system in Chap. 6, practical implementations may take the form of digital emulation. However, gradually a fusion with analog neuromorphic devices and circuits is likely to progress. One of the earliest domains of adoption will likely be specialised applications like medical wearable health monitors, where a dramatic extension of battery life is required. This will make possible continuous operation over durations that were previously challenging (for example, maintenance-free monitoring on the order of months or even years), greatly reducing the burden on users. If patches that adhere to the body and continuously monitor physiological signals over long periods become feasible, they could learn **each user's unique steady-state patterns (homeostasis)** and appropriately detect truly abnormal changes (even if a pattern is generally considered abnormal, it might be normal for that user, and vice versa). During this period, we can also expect the start of efforts toward standardisation of slow electronics. Steps such as establishing common interfaces between devices and creating evaluation methodologies will advance, paving the way for industry to embrace the technology and marking the beginning of a bridge from academic research to industrial application.

By the mid-term (**up to around 2050**), as materials, device technologies, and manufacturing processes mature, the performance of slow electronics devices will have improved greatly, and mass production will likely be possible. In turn, not only wearable health monitors but also a wide range of systems–environmental monitoring sensors, home appliances, social infrastructure, and so on–will routinely incorporate components or modules based on slow electronics principles. Since advanced data analysis and machine learning can be done **locally without relying on the cloud**, privacy-preserving yet highly responsive smart edge devices will become commonplace. This will enable technology to learn even **unconscious habits** that the person himself/herself might not notice. Such capabilities could be used for personal authentication as shown in Chap. 6, but also for **"making visible" tacit skills that are hard to verbalise**–for example, the subtle master techniques of artisans or the super-refined skills of athletes and artists.

Further in the future (**beyond 2050**), slow electronics may have deeply permeated society and could be imparting new values. There are various scenarios for the evolution of computing technology–such as the emergence of quantum computing or strong AI (artificial general intelligence)–but amid these, slow electronics will likely have carved out its own niche as an edge technology that harmonises with humans. In that world, **personal life-assistant devices** might have learned each individual's preferences and long-term habits, and be providing support based on an understanding of "what makes that person unique." Edge devices powered by slow electronics would be trusted as companions that operate at a human pace. Even highly advanced robots, by incorporating slow electronics alongside second-by-second precise control, might be equipped with an interface that allows them to build relationships of trust almost like members of the human society. Slow electronics-driven edge devices would function without relying on high-speed CPUs or constant cloud connectivity, instead being sustained by local slow-electronics circuits. Because their power consumption is extremely low, battery replacements would rarely be needed, and they could even run on harvested ambient energy. As such gadgets appear in everyday life and become recognised as "smart" products that adapt to each individual, **the boundary between humans and technology may dissolve**, with technology becoming an infrastructure that exists as naturally and unobtrusively as air.

We have summarised this envisioned future in Fig. 1.

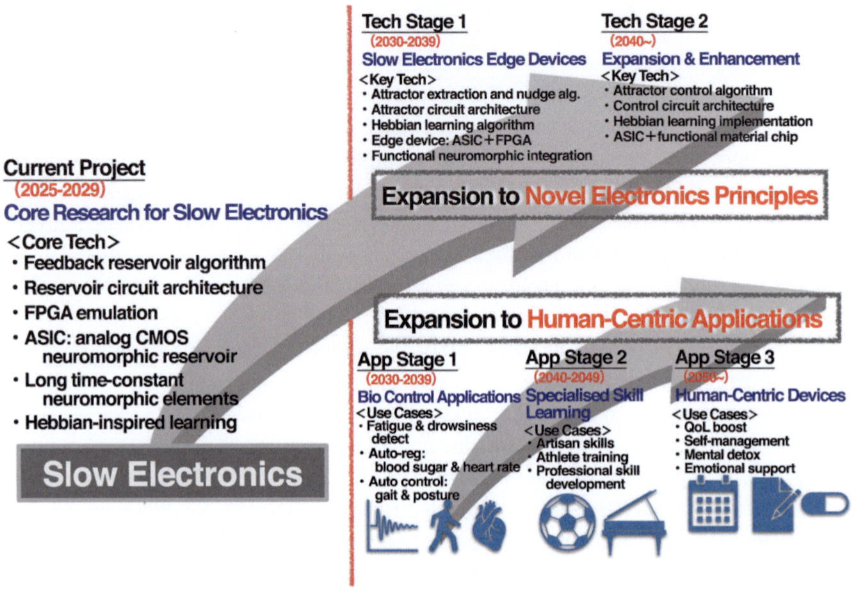

Fig. 1 Roadmap illustrating, in parallel, the progress of slow electronics research and development and its adoption in society. **Top (R&D Progress):** This shows technical advancements in slow electronics. From 2025 to 2030, research progresses on long-timescale electronic devices aimed at reservoir computing implementations, and new artificial neurons and synapses leveraging proton and oxygen-vacancy dynamics are developed. Emulation of these devices on FPGAs also advances, while development of inherently ultra-low-power analog CMOS circuits accelerates. In the 2030s, these foundational technologies become integrated, and implementation in edge devices using ASICs and FPGAs begins in earnest. Research on adaptive circuits following Hebbian learning rules and neuromorphic circuits utilising attractors also progresses, establishing the core technologies of slow electronics. By the 2040s, applied research is expected to accelerate toward embedding these slow-electronics technologies into systems that learn human behaviours and skills. **Bottom (Societal Adoption):** This shows how slow electronics might permeate everyday life. In the 2030s, medical wearables and healthcare sensors become practical, enabling ultra-low-power continuous monitoring of biosignals. In the 2040s, privacy-preserving personal edge AIs emerge that learn individual behavioural traits, making it possible to "visualise" specialised skills and highly advanced expertise. Beyond 2050, personal life-assistant devices that have learned each individual's long-term habits and behaviours become widespread, creating a society where technology closely supports each person. In this way, slow electronics is expected to evolve in parallel on two fronts – advancements in device technology and their applications in society

In the further future, the concept of slow electronics–born from the idea of "adapting to each individual's characteristics"–might even extend to societal infrastructure and industrial systems that were not its initial targets. As technology that learns and adapts to each human's patterns becomes reality, one could imagine **a true smart city** that autonomously and sustainably optimises energy usage and transportation by learning the normal rhythms of an entire city and noticing early signs of deviations (anomalies) from those patterns. What might begin with individual energy-saving home appliances that learn steady environmental changes (temperature, humidity, CO_2 levels, etc.) and autonomously respond to anomalies could spread to factory production lines and agricultural systems. Even in fields traditionally dominated by high-speed control, the introduction of slow electronics concepts (when humans are the focus) is anticipated. Machines and robots–systems not inherently human–might themselves learn their own wear-and-tear and environmental changes to maintain stable operation. By monitoring their "homeostasis" (steady-state conditions) and learning from it, self-maintenance systems for predictive maintenance or smart farming systems that adjust irrigation and lighting to match crop growth rhythms could be implemented. In such a world, technology exists **woven into human life rhythms and societal operations**, and people reap its benefits **without consciously noticing the presence of electronic devices**. In the long run, slow electronics could settle in as more than just a technology–possibly as one of the guiding principles supporting a sustainable, human-centric society. Humans would be surrounded by these "gentle, adaptive systems," benefiting from them almost unconsciously.

Projecting technological progress over the next half-century, one can imagine that the various element technologies related to slow electronics (low-power devices, neuromorphic hardware, edge AI, bio-integrated devices, and so on) will gradually become practical and pervasive in people's lives. Starting with wearable devices and smart homes, by around 2050 we might see full-scale deployment into urban infrastructure and healthcare, and further in the future, a world where human-technology symbiosis is the norm. Of course, the actual future may not unfold exactly as envisioned, but the direction that slow electronics aspires to–**harmony between technology and society**–resonates with the era's calls for sustainability and human-centric values.

5 Impact on Society

Slow electronics promises significant societal benefits, one of the biggest being its contribution to **energy issues**. In today's information society, the electricity consumption of ICT devices is increasing every year and has become impossible to ignore [8]. Slow electronics holds the potential to dramatically reduce the power consumption of electronic devices as a whole. If the enormous number of IoT devices currently in use could operate at the edge on minuscule energy without needing constant cloud connectivity, the load on data centres and communication infrastructure would be alleviated, contributing to a reduction in the ICT sector's power consumption. Moreover, some devices could potentially operate battery-free through energy harvesting, meaning regular battery replacements would no longer be needed, which in turn leads to a reduction in electronic waste (e-waste). This approach would also promote longer device lifespans, helping to realise sustainable electronics that are not simply disposable.

The adoption of slow electronics is also poised to change people's lifestyles. If devices run on ultra-low power and no longer require frequent recharging or battery changes, people will not need to constantly worry about their electronics. For example, wearable sensors and smart home devices could blend seamlessly into daily life, assisting users without drawing attention–realising the concept of "calm technology" where **technology stays quietly in the background**. This is an ideal situation where technology does not excessively demand user attention and allows people to concentrate on their own activities. In fact, such a gentle, user-attuned approach to technology was proposed by Hallnäs and Redström as a design philosophy called "slow technology" [9]. Slow electronics embodies this philosophy at the hardware level, enabling **a way of living with technology that does not place burdens on the user**.

Furthermore, handling data locally means that **privacy is preserved**, enabling a society in which people can use technology with peace of mind. As mentioned earlier, if data is processed on an edge device, the risk of personal information leaking out is greatly reduced. Devices employing slow electronics are not only energy-efficient but also inherently oriented toward on-device processing, making them reliable, privacy-protecting smart devices. For instance, a health-monitoring wearable could analyse a user's biometric data internally and only send out summary information when necessary, freeing the user from the feeling of being constantly surveilled.

The notion of "slowness" could even influence our overall relationship with technology. Instead of perpetually seeking "faster and newer," being willing to accept a slower pace when appropriate may help ease the stress of digital life. If slow electronics becomes widespread and devices last longer without needing frequent replacements, consumers can continue using their favourite gadgets for many years. This shift would encourage us to **rethink the traditional "mass production, mass consumption" lifestyle**, possibly steering us toward a more sustainable and human-centric way of living. A society in which slow electronics harmonises with human life could ultimately provide people with greater peace of mind and lay the groundwork for a truly rich and fulfilling life.

6 A Day in the Life of a Future Family

Slow electronics will bring about a society where technology and humans truly coexist. In this future, electronic devices quietly support our daily lives, working in the background to enhance comfort, security, and well-being–without demanding attention. This aligns with the vision of Society 5.0, a human-centred, sustainable society where the digital and physical worlds merge seamlessly, [10]. Slow electronics will serve as the foundation for this vision, enabling a world where **technology feels less like a tool and more like a personal companion**.

But what does this actually look like? Let's follow a **family in 2050**, living in the suburbs, where slow electronics is an invisible but ever-present force shaping their lives.

6.1 *Morning: A Personalised Start to the Day*

As the sun rises, the **family's home, powered by slow electronics**, has already started adjusting the environment based on each member's needs. The mother, an office worker, usually wakes up to a gentle light simulation mimicking a natural sunrise. But today, her stress levels were slightly elevated overnight. **Her personal AI assistant**, embedded in her bedside device, has quietly shifted her wake-up time by ten minutes and rescheduled her commute for a slightly later train to allow for a more relaxed morning.

In the kitchen, the eldest son, who works as a caregiver, is already up. His smartwatch, equipped with slow electronics, **has learned his caffeine tolerance over time** and knows today is one of those mornings where he needs a strong cup. The coffee machine, acting as his **personal barista**, is brewing his preferred blend–extra strong, no sugar.

Meanwhile, the daughter, a primary school teacher, groggily stares at her smart mirror while brushing her hair. The mirror, which **analyses facial expressions in real-time**, detects mild drowsiness and stress and automatically plays her favourite upbeat morning playlist.

The father, who is managing **high blood sugar levels**, receives a gentle nudge from his **wearable health assistant**. "Good morning! Time for a short walk after breakfast," it suggests, based on real-time glucose levels. He sighs but chuckles–**his device knows him too well**.

6.2 *Commuting: An Extra Pair of Eyes*

The mother, now feeling refreshed, heads to the station. Her **smart assistant**, embedded in her wearable device, has checked live train congestion data and advised her to board the train with the least crowding, making her commute as comfortable as possible **without her even realising it**.

On his way to work, the eldest son's **navigation assistant** detects road congestion and suggests an alternative route. "Take the river path today. Less traffic, and it's good for your stress levels," it advises. He follows the suggestion, grateful for a moment of quiet before a busy day at the care facility.

6.3 *At Work: The Invisible Helping Hand*

At the **hospital**, the father arrives for his routine blood test. His **wearable health tracker**, which continuously learns his unique physiological patterns, has already flagged minor fluctuations and updated his medical records. The doctor, already aware of the changes, adjusts his prescription on the spot. "Your assistant is sharp," the doctor laughs. "It caught this before you even noticed."

At the **nursing care facility**, the eldest son is making his rounds. A robotic caregiver equipped with slow electronics approaches him. "Mr. Ohtani seems quieter than usual. His walking speed is slightly reduced–should we check on him?" it asks. Because the AI has been learning Mr. Ohtani's **unique habits**, it detected this subtle change before anyone else did. The eldest son nods and heads over, **preventing a potential health issue** before it worsens.

At the **primary school**, the daughter is preparing a lesson. Her **personal teaching assistant**, embedded in her classroom tablet, suggests grouping students based on their learning patterns. The AI has been silently observing students' **focus levels, emotional states, and interactions** to determine the best group dynamics. "Today's groupings look promising," she thinks, as she watches shy students paired with encouraging classmates.

6.4 Evening: A Seamless Transition to Home Life

After a long day, everyone returns home, and their personal slow-electronics devices synchronise their health, stress, and fatigue levels with the home system.

The father, slightly tired from his hospital visit, finds the house has adjusted the lighting and temperature to help him relax. His wearable device has also planned an ideal post-dinner walk, considering his glucose levels and fatigue.

The mother, still feeling the stress from work, is greeted with her favourite herbal tea, already prepared by the kitchen assistant. The eldest son unwinds on the sofa, where the system has selected a relaxing podcast, while the daughter reviews the day's classroom activities through an AI-generated summary that highlights key student interactions.

Finally, as they all wind down, the **slow-electronics-powered home adjusts itself**–lights dim subtly, the temperature drops slightly for optimal sleep, and the family's wearables analyse sleep quality in real-time, ensuring they wake up refreshed for another day.

6.5 Technology That Feels Like Family

In this future, **slow electronics does not demand attention**–it simply understands, anticipates, and adapts. Devices are no longer mere tools; they have become **trusted companions**, deeply attuned to each individual's health, habits, and personal well-being. Instead of a world where technology feels intrusive, slow electronics has created a reality where machines are more like caring family members–a personal doctor, a private secretary, an intuitive advisor–**always present, yet never overwhelming**. This is not just a vision of the future–it is the world that slow electronics is bringing to life.

7 Final Remarks

We have sketched the future vision of slow electronics and its impact on society. Real-world technological development may encounter difficulties at times, but as highlighted in this book, a solid foundation is already being steadily built in each relevant field. Once slowly-operating electronic devices, architectures, and algorithms are integrated, the dream-like scenarios described in this chapter will one day no longer be mere dreams.

Slow electronics is not just a new technology, but offers a perspective complementary to the traditional value of "**speed = superiority**" by presenting a **new information-processing framework that adapts to long timescales**, thereby reconstructing the relationship between humans and technology. We have great hopes for the future. To an electronics industry that has hitherto been "all about speed," slow electronics will propose a new value of "being kind to humans." It is, in a sense, a movement to re-examine the relationship between people and technology and to return to technology's original goal of pursuing human happiness. The future opened up by gentle, human-attuned electronics overlaps with the future that we truly desire.

Finally, as we close this book, I want to emphasise that progress in this field will require **interdisciplinary fusion and open-minded thinking**. Only by bringing together diverse knowledge–electronics, computer science, mathematics, physics, materials science, and even psychology and medicine–can truly human-friendly technology take shape. I sincerely hope that each of you, in your respective positions, will exercise your creativity and contribute to the development of slow electronics. And someday, let "being slow" become the hallmark of cutting-edge technology, as we realise a future where humans coexist with technology in harmony. The journey of slow electronics has only just begun. Embracing the boundless possibilities that lie ahead, we conclude this book with endless hope for the future.

Competing Interests
This study was funded by Japan Science and Technology Agency (JST) CREST Grant No. JPMJCR19K.
The author has no conflicts of interest to declare that are relevant to the content of this chapter.

References

1. G. Tanaka, T. Matsumori, H. Yoshida, K. Aihara, Phys. Rev. Res. **4**, L032014 (2022). https://doi.org/10.1103/PhysRevResearch.4.L032014
2. H. Inoue, H. Tamura, A. Kitoh, X. Chen, Z. Byambadorj, T. Yajima, Y. Hotta, T. Iizuka, G. Tanaka, I.H. Inoue, Adv. Mater. **37**, 2407326 (2024). https://doi.org/10.1002/adma.202407326
3. S.P. Pati, Y. Geng, S. Hamasuna, K. Fujiwara, T. Iizuka, H. Inoue, I. Inoue, T. Yajima, Commun. Mater. **5**, 1 (2024). https://doi.org/10.1038/s43246-024-00621-1
4. X. Chen, T. Yajima, I.H. Inoue, T. Iizuka, Jpn. J. Appl. Phys. **61**, SC1051 (2022). https://doi.org/10.35848/1347-4065/ac43e4
5. X. Chen, Z. Byambadorj, T. Yajima, H. Inoue, I.H. Inoue, T. Iizuka, Appl. Phys. Lett. **122**, 074102 (2023). https://doi.org/10.1063/5.0136627
6. H. Tamura, G. Tanaka, Neural Netw. **143**, 550 (2021). https://doi.org/10.1016/j.neunet.2021.06.031
7. K. Koyama, H. Ando, K. Fujiwara, AIP Adv. **13**, 045317 (2023). https://doi.org/10.1063/5.0124625
8. A. Andrae, T. Edler, Challenges **6**, 117 (2015). https://doi.org/10.3390/challe6010117
9. L. Hallnäs, J. Redström, Pers. Ubiquitous Comput. **5**, 201 (2001). https://doi.org/10.1007/pl00000019
10. A. Deguchi, C. Hirai, H. Matsuoka, T. Nakano, K. Oshima, M. Tai, S. Tani, in *Society 5.0* (Springer, Singapore, 2020), pp. 1–23. https://doi.org/10.1007/978-981-15-2989-4_1

MIX
Papier aus verantwortungsvollen Quellen
Paper from responsible sources
FSC® C105338

If you have any concerns about our products,
you can contact us on
ProductSafety@springernature.com

In case Publisher is established outside the EU,
the EU authorized representative is:
Springer Nature Customer Service Center GmbH
Europaplatz 3, 69115 Heidelberg, Germany

Printed by Libri Plureos GmbH
in Hamburg, Germany